PRAISE OF CHROMOSOME "FOLLY"

Confessions of an Untamed Molecular Structure

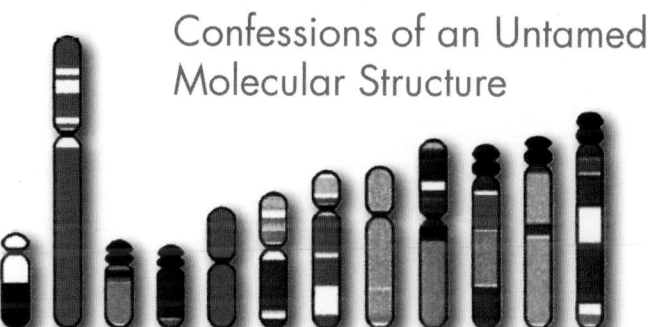

PRAISE OF CHROMOSOME "FOLLY"

Confessions of an Untamed Molecular Structure

Antonio Lima-de-Faria
Lund University, Sweden

 **World Scientific**

NEW JERSEY · LONDON · SINGAPORE · BEIJING · SHANGHAI · HONG KONG · TAIPEI · CHENNAI

Published by

World Scientific Publishing Co. Pte. Ltd.

5 Toh Tuck Link, Singapore 596224

USA office: 27 Warren Street, Suite 401-402, Hackensack, NJ 07601

UK office: 57 Shelton Street, Covent Garden, London WC2H 9HE

Library of Congress Cataloging-in-Publication Data
Lima-de-Faria, A.
  Praise of chromosome "folly" : confessions of an untamed molecular structure /
Antonio Lima-de-Faria.
    p. ; cm.
  Includes bibliographical references and index.
  ISBN-13: 978-981-281-479-1 (hardcover : alk. paper)
  ISBN-10: 981-281-479-5 (hardcover : alk. paper)
  ISBN-13: 978-981-281-094-6 (pbk. : alk. paper)
  ISBN-10: 981-281-094-3 (pbk. : alk. paper)
  1. Chromosomes. 2. Mutation (Biology) 3. Genetics. I. Title.
  [DNLM: 1. Chromosomes. 2. Evolution. 3. Mutation. 4. Selection (Genetics)
QU 470 L732p 2008]
  QH600.L555 2008
  572.8'7--dc22

                                                        2008023227

British Library Cataloguing-in-Publication Data
A catalogue record for this book is available from the British Library.

Typeset by Stallion Press
Email: enquiries@stallionpress.com

Printed by FuIsland Offset Printing (S) Pte Ltd, Singapore

To the Reader

It may be inappropriate to start by addressing the Reader, but the title of this book is so unconventional, that some form of explanation may be called for.

Why treat such a serious subject, as the molecular organization of the chromosome, in the form of a satire? The reason may not seem obvious but is simple.

On one hand, two World Wars (1914–1918 and 1939–1945) were followed by a long period of "Cold War" (1946–1991) characterized by a severe economic and political confrontation that limited intellectual manoeuvre.

On the other hand, globalization has led to an enormous economic and technological growth that has allowed the realization of large scale projects. However, huge profits became mandatory and these demanded the canalization of scientific research into specific alleys.

Following in the steps of the multinationals, the leading scientific publishers, acquired the small ones and in turn fused with each other, extinguishing plurality. The same happened to scientific journals. Publication has become reduced to a few leading journals. Articles in minor ones are not being cited.

The rarefaction of alternatives has become so acute that outstanding colleagues, in the U.S.A. and the United Kingdom, being aware of this situation, proposed several modifications that have had difficulties to materialize. Lawrence (2003) stated: "*Nature* now receives around 9,000 manuscripts a year (double that of 10 years ago) and has to reject about 95% of biomedical papers" and after enumerating the many publishing difficulties faced at present by scientists he added: "These forces all combine to create an anti-scientific culture."

The situation has been getting tighter, as the interests of governments became involved. In face of this rapid development the New Scientist felt obliged to write an Editorial with the title: "Dirty tricks. If science doesn't suit your political viewpoint, suppress it" (2007). The article refers to the Atmosphere of Pressure, a report published in 2007 by the Union of Concerned Scientists and the Government Accountability Project. The Editorial concludes: "We all know that governments take scientific findings and use them or ignore them according to their ideological goals". The same tends to be done by scientists whose research is supported by government funds.

Several factors seem to have contributed to create in Western society an ideological vacuum which imposes a single voice on many scientific and cultural issues.

This book is for those who have perceived that the scientific endeavour is an ever changing process in which the explanations available at a given time only lead to a partial understanding of the known phenomena. It is directed to inquiring minds who have harbored doubts about the correctness of prevailing monolithic ideas and have been looking for alternative explanations.

Dear Reader, if your mind is locked in a conventional attitude towards science, please step aside. Moreover, if you have an escapist attitude directed to obscurantism, this is not either the book that you are looking for.

References

Lawrence PA. (2003) The politics of publication. Authors, reviewers and editors must act to protect the quality of research. *Nature* **422**: 259–261.

Editorial. (2007) Dirty tricks. If science doesn't suit your political viewpoint, suppress it. *New Sci* (Feb. 3rd 2007): 5.

Contents

Introductory Note

An effort was made to use as few technical terms as possible throughout the text. Moreover, when they are introduced in a sentence they are usually explained right away. *Cytology* is the study of the cell's structure and function, whereas *genetics* deals with hereditary transmission. In the 1960s biochemical methods combined with electron microscopy and radioisotope labelling, brought cytology and genetics to a common ground creating a new discipline: molecular cyto-genetics. This became, two decades later, generally known as molecular biology.

CHROMOSOME — The most important of all cell organelles due to being the main agent of genetic information. 1. In organisms that lack a nucleus in their cells (called prokaryotes), such as the bacteria, the chromosome is a circular DNA molecule containing the genetic information and is free in the cytoplasm. Part of its DNA is associated with a few proteins. 2. In organisms with cells having a nucleus (eukaryotes), the chromosome is one of the threadlike structures consisting of scaffolding proteins, basic proteins (histones) and DNA. This last macromolecule carries genetic information arranged in a linear sequence. In a eukaryotic cell besides the nucleus there are other organelles, such as mitochondria and chloroplasts, that have their own chromosomes consisting of circular DNA which also carry genetic information. They are usually of smaller dimensions. In all cases the genetic information is decided by the order of the bases along the DNA.

DNA transfers its genetic information by building a messenger RNA molecule complementary to its base sequence (transcription). This in turn leads to the synthesis of a protein with a specific configuration (translation). The situation may be reversed when RNA is transcribed into a DNA molecule, with the help of the enzyme reverse transcriptase, an event which occurs in some viruses.

The Source of "Folly" and the Reason for Confessions

1

I am an Inconspicuous and Unattractive Creature Painted with Lipstick

I am so inconspicuous that no one can see me with the naked eye. My length is usually between 1 and 10 thousand parts of a millimeter.

For millennia no one knew of my existence. This is why I am not included in the Ancient Egyptian paintings that depict plants and animals. I am not part of Michelangelo's frescoes in the Sistine Chapel that represent the birth of the world. And, in modern times, there is no indication that Picasso included me in any of his 20,000 works.

No one knew of my existence until microscopes were developed allowing an object to be magnified as many as 1,000 times.

But it was not so easy. The organs and tissues, of plants and animals, had to be separated into that unit of life that is the cell. Subsequently, only by breaking the cell, could I be found, well encased in its bosom, in company with the other minute molecular structures (Fig. 1.1).

Locked in this prison, with thick membranes and walls surrounding me everywhere, I looked pale, transparent and hardly distinguishable from my surroundings.

Ladies, of the upper classes, had been colouring their cheeks and lips with a dye called carmine, extracted from the body of a

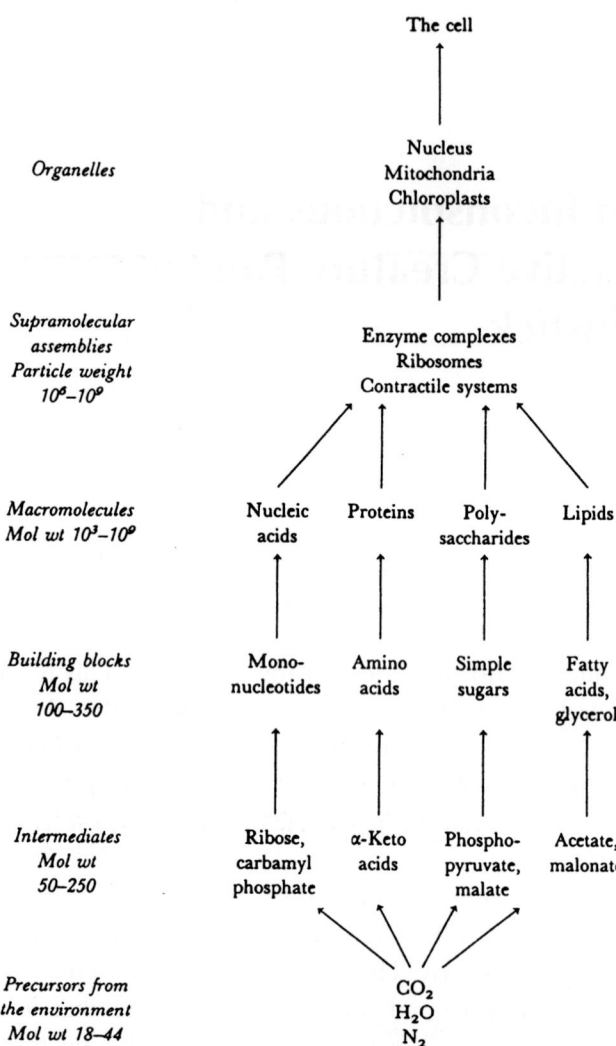

Fig. 1.1   Successive stages in the formation of the cell.

Starting with precursors from the environment, larger molecules were built. These building blocks led to the formation of the macromolecules: nucleic acids, proteins, polysaccharides and lipids; which in turn resulted in supramolecular assemblies such as the ribosomes, where protein synthesis occurs. The next step was the formation of independent organelles until one finally reached the cell. The nucleus, mitochondria and chloroplasts contain chromosomes. Those of mitochondria and chloroplasts are rudimentary structures. Most of the organism's genetic information is concentrated in the large and complex chromosomes of the nucleus.

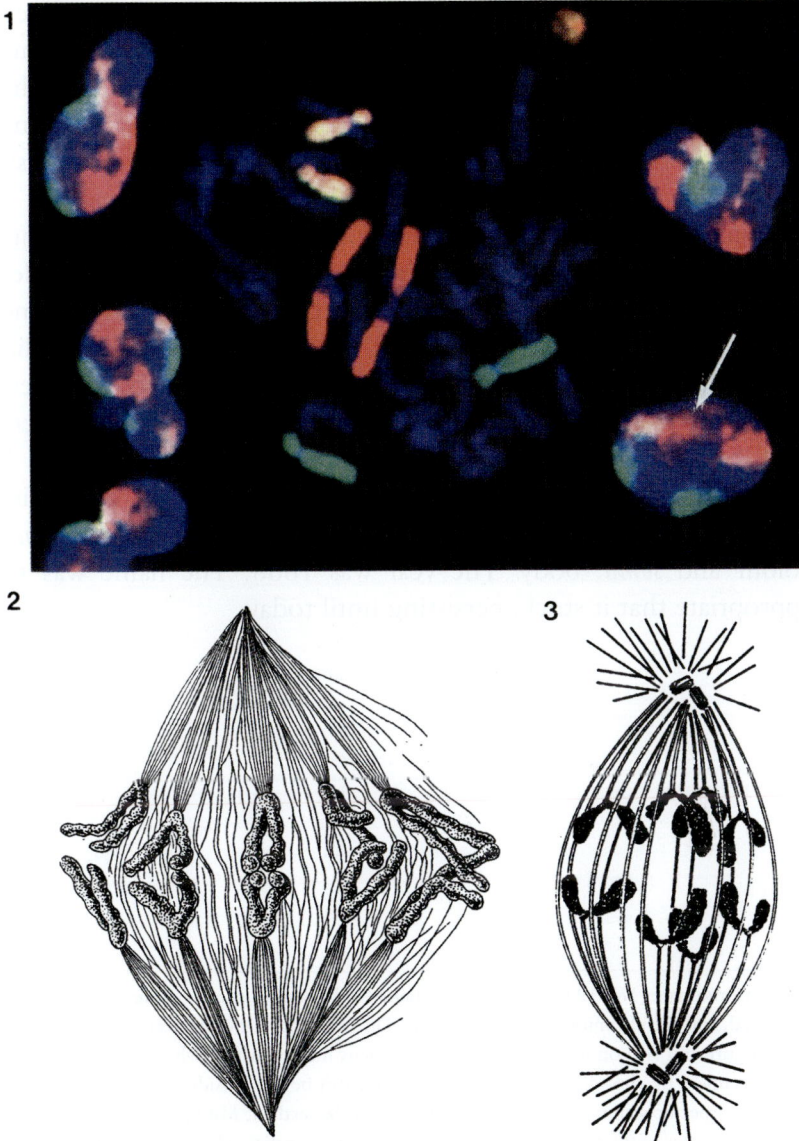

Fig. 1.2   Painted chromosomes and their movements during cell division.

1. In the central part of the figure are the 46 chromosomes (23 from the father and 23 from the mother) of the human complement at metaphase of mitosis, a stage at which the chromosomes are highly contracted and soon separate into daughter cells. On purpose most chromosomes are stained a faint violet being hardly seen. By using novel techniques the two

scale-insect, which lived on the cactuses of tropical America. When a drop of carmine was placed over the disrupted cell I easily became coloured a bright red and could be well recognized. Catherine the Great, Empress of Russia (1729–1796), as well as Marquise Jeanne de Pompadour (1721–1764) already used this dye to embellish their faces.

During the same time, naturalists who were looking for plants among the snowy peaks of the Alps, discovered the beautiful violet flowers of *Gentiana* . An extract from these flowers, covered me with an impressive purple colour. I was then dressed in splendid red, violet, green and, even later, with shining fluorescent dyes. I appeared as colourful as a tropical bird with its brilliant feathers (Fig. 1.2).

I liked these fancy dresses so much, that the German cytologist W. Waldeyer called me *chromosoma*, from the Greek words *chroma*, colour and *soma*, body. The year was 1888. The name was so appropriate that it stuck, persisting until today.

---

small chromosomes No. 12 were selectively stained light orange, the two long chromosomes No. 1 red, and the chromosomes No. 4 green. Three resting nuclei are seen on the left side of the figure and two on its right side. In the nucleus marked with a white arrow, the territories occupied by chromosomes Nos. 12, 1 and 4 can be seen inside the resting nucleus, due to the specific staining. Image by courtesy of H. Scherthan, Munich, FRG.

2. Chromosomes of the plant *Lilium* dividing in a pollen mother cell. The attachment of the chromosomes to the spindle fibers takes place in the median region where the centromere is located. The spindle fibers move the daughter chromosomes to opposite poles. In this species, as in other plants, there are no asters and no centrioles.

3. Cell division in an animal cell. In animals two other types of organelles have been added to the moving apparatus. They are the asters (radiating fibers located at the opposite poles) and the centrioles (a pair of organelles from which the asters radiate).

2

They Say that I Resemble a Sausage

B efore I was painted with these dyes they impregnated me with alcohol, formalin and other chemicals to preserve my body. Actually I was embalmed like an Egyptian mummy to be kept for thousands of years in a coffin. My interior molecular structure was not revealed by this drastic treatment and my outer surface was as smooth as that of the mummy's coffin. I actually looked like an elongated cylinder when I was most conspicuous during the cell division (Fig. 1.2).

If I may use gastronomic terms I may be compared to a sausage having, like most of them, a constriction in the middle of the body or in its vicinity which harbours the centromere, a region participating actively in chromosome movements. This constriction separates my body into two parts which cytologists call "arms", but I have neither arms nor legs. I am even more deceptive, because whereas a sausage has a skin surrounding it, surprisingly I do not have an outer membrane. Cytologists thought, for many years, that I had to have a solid shell, since my body appeared so well delimited at most stages. Finally the idea was abandoned, because no membrane could be found. In this respect I am quite diifferent from other cell organelles. My boundaries are a product of my internal molecular edifice.

3

The Cell is My Castle and Prison But I May Swim and Dance like an Odalisque in a Harem

The cell is my castle but at the same time is undoubtedly my prison, since I cannot get out of it and I never lived anywhere else. I know that I am condemned to this perpetual confinement. But the view is not gloomy. Inside its walls I have a lot of fun because I am in a constant state of amusement.

Inside the cell there is a swimming pool that happens to be like a sphere. They call it the nucleus. Most of the time I am enclosed in this sphere and here I swim permanently in all directions like a fish in a fishbowl (Fig. 1.3). I was the main star in several films in which I can be seen moving elegantly, like an odalisque in an oriental harem, in what they call the nuclear sap.

But nothing is static in living processes. The walls of this nice and warm aquarium suddenly break down and I find myself ejected into a disagreeable cytoplasm much more viscous than my previous dwelling. There is no alternative but to get rid of this situation as soon as possible. I start to dance together with my other comrades, moving in a well planned choreography (Page and Hawley, 2003). We start by attaching ourselves to ribbons and move backwards and forwards in the middle of the cell. Like in a ballet, following the sounds of a Viennese waltz, we dance and finally move in opposite

Fig. 1.3   Chromosomes are prisoners of the cell as much as fish are of an aquarium. Chromosomes thrive in their own environment but cannot function outside it. "Still life with goldfishes" 1911 by the French painter Henri Matisse (1869-1954).

directions reaching the corners of the cell. Then, the music stops, and the elements of the nuclear envelope that had been left in the cytoplasm waiting for the play to finish come again from every side. They converge on us and rebuild rapidly a new nucleus. Suddenly, I am back again in my private aquarium, becoming protected from all those convulsive chemical reactions that are going on in the rest

Fig. 1.4   The Aquarium as a symbol of life's barriers.

No other painter seems to have been so fascinated by this motif as H. Matisse. He depicted it at least eight times both in oils and drawings. In two cases Matisse placed women absorbed in deep thought, contemplating the fishes trapped in the bowl. "Woman before an Aquarium" was painted in 1921.

of the cell. Nothing like the quiet peace of the nucleus, where I can copy myself without being disturbed and prepare my progeny in splendid isolation. The advantages are many, but the limitations imposed on my behaviour are equally impressive.

If I assume the form of a virus, I may escape from the cell for a time, acquiring in many cases a crystal structure. But this escape is only temporary. To reproduce as a virus, I must go back to the cell and only by using its molecular apparatus may I create a new progeny (Luria *et al.*, 1978). Under all conditions the return to the cellular prison is obligatory (Fig. 1.4).

4

The Striptease Show — How I Dress and Undress at Every Cell Division

There is another side of my behaviour that is equally foolish.

Indiscreet film directors, by using cine-cameras fixed on the top of microscopes, have documented my private life. In their film sequences, made on plant and animal cells, I appear to change my dress all the time (Bajer, 1957).

When I swim inside the nucleus, I do not even have a bathing suit. No membrane covers me as it does in the case of other cell organelles.

During cell divison, I actually perform a striptease show, but I do it in reverse order. I start by being naked and dress successively. Then I undress, finishing by becoming like Adam and Eve without a fig leaf.

Actually, I am as tricky as a clown during a circus performance. From the beginning my suit gets loaded with embroidered small decorations, called chromomeres, as my threads are knotted into minor spirals. Without the public noticing it I display another dress within minutes. Also I extend and contract my body like an acrobat (Fig. 1.5). As cell division progresses, I am ejected out of the fishbowl into the cold and inhospitable cytoplasm. Happy but shivering, I protect myself by contracting heavily and looking as being

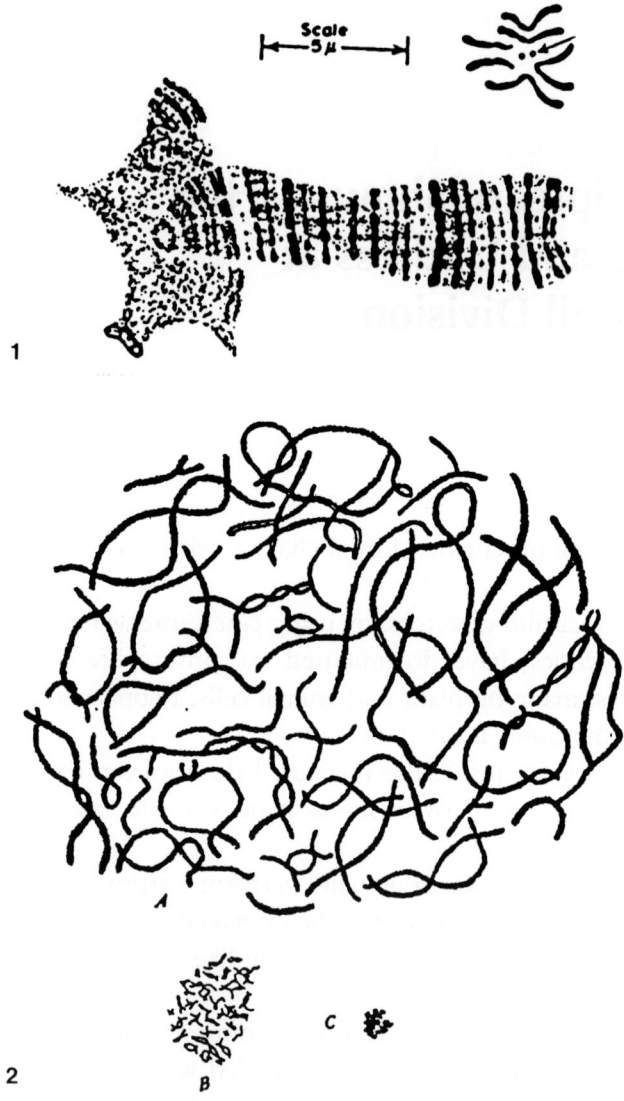

Fig. 1.5 The same chromosome drastically changes its size and configuration during the organism's development.

1. The chromosome 4 of the fly *D. melanogaster*, as it appears in the cells of the salivary glands (16 microns long) and in gonial cells (upper right, and indicated by arrow) where it is only a fraction of a micron.

2. Chromosomes of the shark *Pristiurus* at different periods of egg development drawn to the same scale. A, period of maximal size; B, later period; C, at the close of ovarian life.

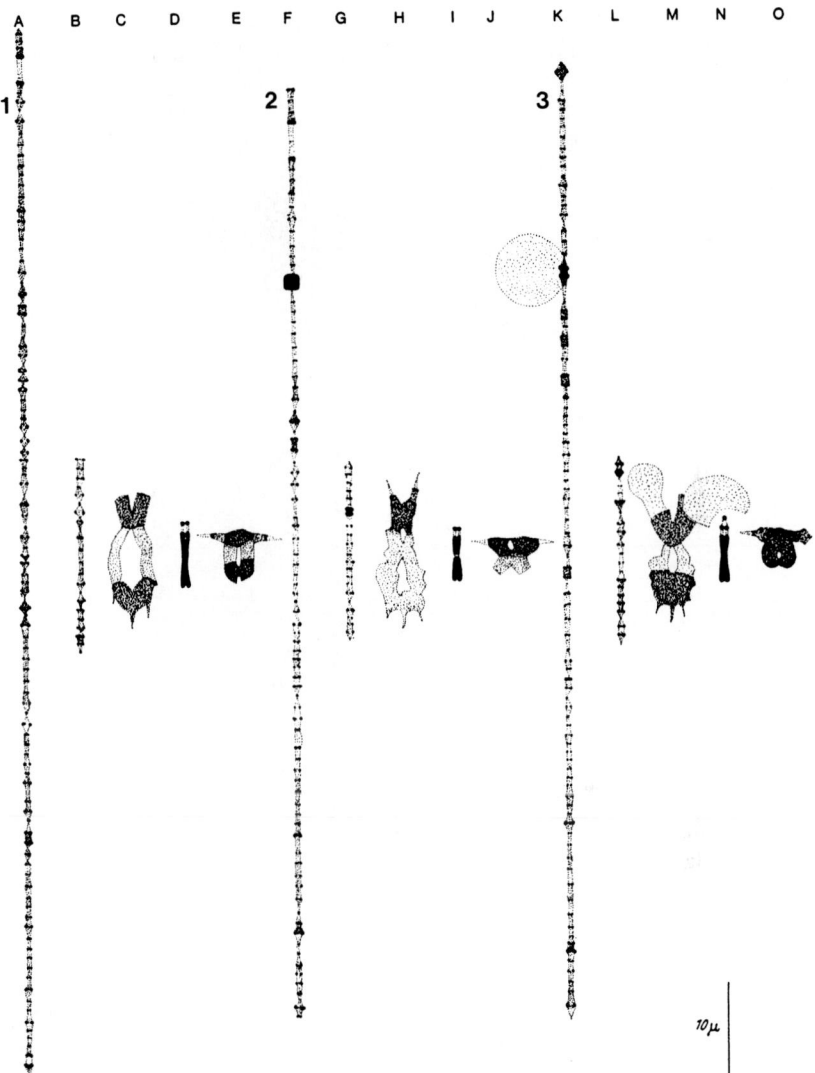

Fig. 1.6   The same chromosome changes its length from 107 to 7 microns — a 15 times decrease in size.

Ornithogalum virens is a plant with only 3 chromosomes. These can be distinguished at the various stages of cell division and in different tissues. Chromosome I (A-E) chromosome II (F-J) and chromosome III (K-O). At the first prophase of meiosis each chromosome is highly distended displaying over 100 chromomeres along its body (A, F, K). At the second pollen prophase the chromosomes are now much shorter but still show a distinct chromomere

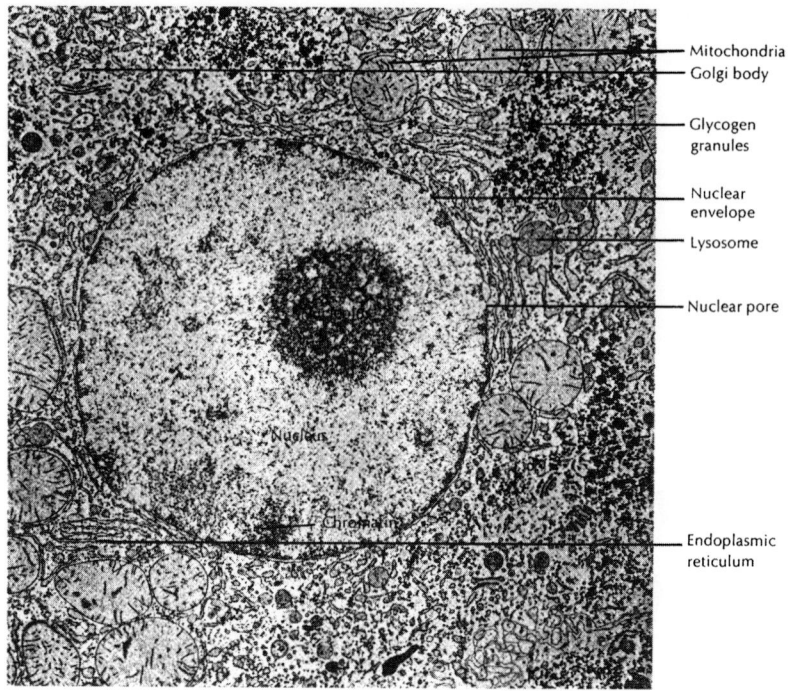

Mitochondria
Golgi body

Glycogen granules

Nuclear envelope

Lysosome

Nuclear pore

Endoplasmic reticulum

Fig. 1.7   Chromosomes can make themselves invisible.

An electron micrograph of a liver cell from a rat. The large resting nucleus has in the center a distinct dark nucleolus (see Fig. 1.8), but the chromosomes of the rat are not visible. This is a regular feature. Chromosomes become invisible in the nuclei of differentiated cells. Compare with resting nuclei in Fig. 1.2 in which specific chromosomes were made visible at this stage, by using a special staining technique.

wrapped in a fur coat (Fig. 1.6). Then as I reach the spindle poles and movement stops, I may strip again. But I go one step further, making myself invisible, in a way better than any magician could

---

pattern (B, G, L). During late prophase of meiosis (diakinesis) (C, H, M) the paired chromosomes start to separate and have contracted heavily. At metaphase of mitosis they are small rods (D, I, N). Finally at metaphase I of meiosis they become highly condensed (E, J, O). In future cell divisions the three chromosomes revert to their large sizes, the restructuring cycle being repeated again and again. Meiosis is the cell division that leads to the formation of sex cells (see Fig. 1.8) and mitosis is the division that occurs in most tissues of the body. Prophase and metaphase are early and later stages of these processes.

do. During the so called "resting nucleus" no one can see me although I am there, ready to start the next performance, as the curtain goes up giving the sign for the start of the next cell division (Fig. 1.7).

It is remarkable that the same type of show — in which I am a dancer, a clown, an acrobat and a magician — has been displayed since the dawn of the protozoan cell, in an untold and astronomical number of times. Like a musical, that runs for months on Broadway, this show has gone on for millions of years.

5

I Have Created My Own Private World Full of Tricks, Back Door Exits and Novel Solutions — I am an Untamed Innovator

Be independent. Never lean on the opinion or behaviour of others — that is my motto or better, the principle that I have always followed. How could I survive throughout billions and billions of cell divisions, copying myself into an untold number of descendants, and changing place and dress at every moment without dying at an early age. I would long ago have lain on the ground full of scars and amputated arms, I simply would not be around.

Above all I had to rely on my power, but to survive I had to invent a number of tricks, back door exits and create novel solutions.

The tricks are many. To mention only a few. In the early days of evolution, when I resided in the bacteria, I needed no ends, because I was a circular structure. But as the nucleus was formed and I became a linear rod, I had to cope with a novel situation, what could I do with my new ends? I started by placing at my terminal regions well defined caps that locked them (the telomeres). To move more comfortably to the spindle poles during cell division I devised a region of attachment to the large ribbons of the spindle

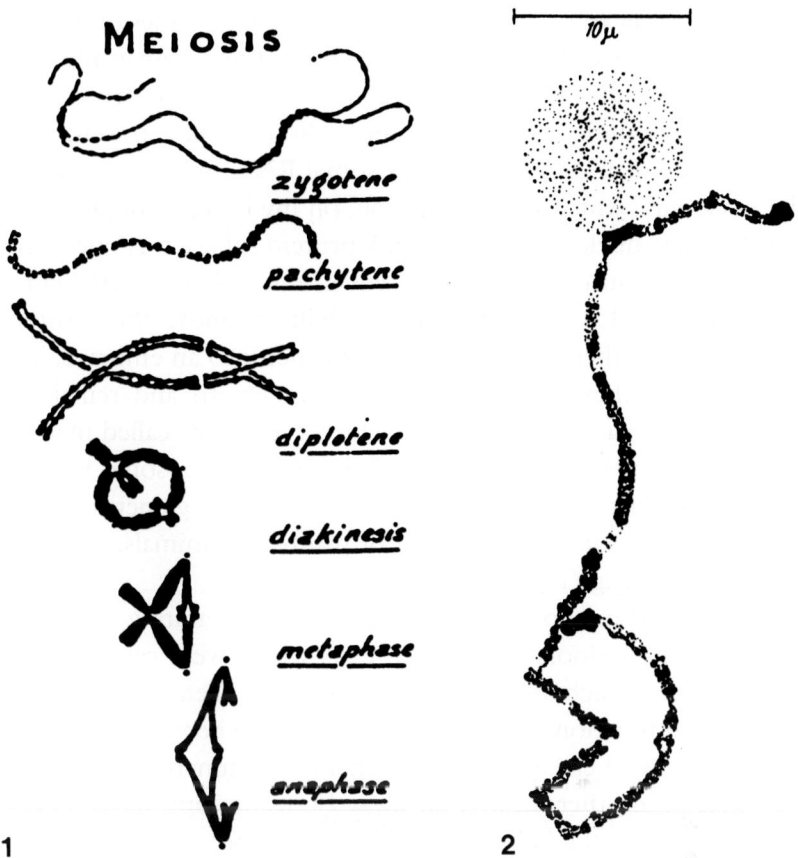

Fig. 1.8   Pairing and contracting of chromosomes and the spherical shape of the nucleolus.

1. Meiosis is a specialized form of cell division during which the chromosome number of the individual is reduced to half resulting in the formation of germ cells. In its initial phase the chromosomes of the mother and those of the father come close to each other and pair (stages zygotene and pachytene). Initially the chromosomes are very long and display a series of minute bodies called chromomeres. During the following stages the paired chromosomes exchange segments an event called crossing over (diplotene). Later they move out of the nucleus and contract heavily during metaphase. At anaphase they start moving to opposite spindle poles.

2. The chromosome of rye which bears the nucleolus at meiosis. In most plants and animals this has the shape of a large sphere, consisting of RNA and protein, and is attached to the chromosome body at a specific site. The rest of the chromosome displays its chromomere pattern.

making it easier to reach the correct site at the extremes of the cell. Such a region is called a centromere and did not exist in the early periods of my adventurous life. Like the scarab beetle that collects a large ball of dung and carries it throughout its daily travels, I too created an equally large ball of ribosomal RNA and protein (called the nucleolus) which I carry with me, on my back, throughout my travels in the nucleus. Such an RNA-protein sphere was one of my inventions because it was not present in the earlier bacterial chromosome (Fig. 1.8). Also I have the ability to move the chromosomes of the father and those of the mother into an embrace. The result is that they pair along their entire length and remain so throughout a part of this division which cytologists called meiosis. The name derives from the halving of the chromosome number occurring in the ensuing cells. These become the gametes participating in the sexual reproduction of plants and animals, such as a sperm and an egg (Fig. 1.8).

How could my body be defended from its assaulting enemies if it had not back doors and unseen exits which were camouflaged. I carry genes which are kept under cover, not seen easily, because they are not functional but which at a signal can become fully active (pseudogenes). I can also suddenly change the function of a region into that of another located far away by using long range effects carried out with the help of small RNAs and proteins (Carrington and Ambros, 2003; Mello and Conte, 2004).

The most critical part of my existence was during the permanent copying of my structure. Temperature and environmental fluctuations, created a series of deviations that disturbed my integrity and coherence. But I was on my guard. I called on specific proteins to help me to recognize any errors that I committed and to repair the original message correctly (Wood, 1996). That was undoubtedly one of my most impressive achievements. Without this mechanism there would never have been a chromosome that would have maintained its existence on the planet for so long time.

6

My Origins Were Humble — The Antithetical Nature of Matter Left a Mark on My Construction

The study of the origin and formation of the universe is the domain of Astrophysics. The picture that physicists have so far attained is one in which stars and galaxies are in a super dynamic balance. Stars explode into myriads of other stars whereas others condense into "black holes" of a colossally dense mass. The outer universe is an untamed world. Everywhere it is creating and annihilating, producing light and darkness, at tremendous speeds (Burrows, 2000; Hillebrandt *et al.*, 2006). Moreover, the elementary particles that constitute matter are counterbalanced by the existence of antiparticles that have been successively discovered. These are the antielectrons (positrons), the antiprotons and the antineutrons (Alfvén, 1966; Pagels, 1982; Quinn and Witherell, 1998).

Matter and antimatter are an essential part of this antithetical duality (Fig. 1.9).

It is from this contradictory world that the chromosome emerged at a time, and at a place, where this convulsion seemed to have been settling down to a calmer pace.

Like the mythological god Atlas carrying on his curved shoulders the tremendous weight of the world, I carry in my body all the contradictions and constraints of the early universe.

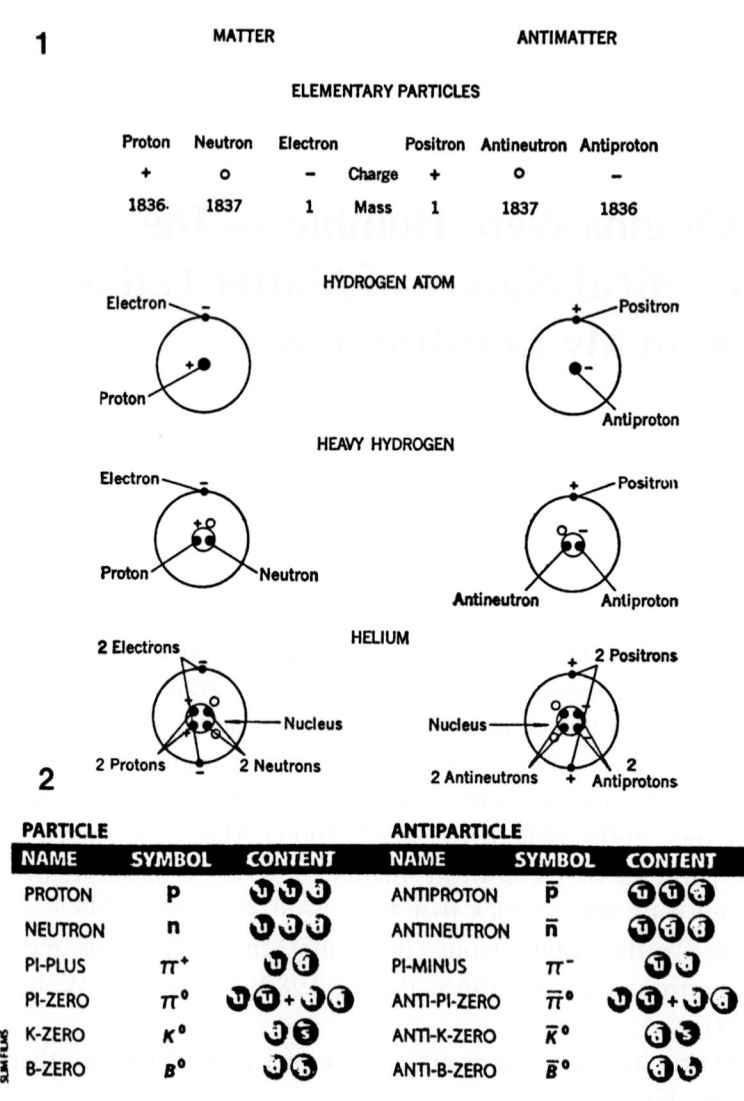

Fig. 1.9 Matter and antimatter are components of the atomic structure.

1. The particles and antiparticles, which build matter and antimatter as known in 1966. Antimatter consists of positrons (antielectrons), antiprotons, antineutrons and other particles.

2. Presently known particles and antiparticles showing their quark content.

No new elementary particles, new protons or electrons, even new atoms, were created when I was born. The chemical analysis of stars, by spectrography and other methods, reveals that the chemical elements found in the whole universe are the same that constitute living organisms (Woosley and Janka, 2005).

The universe did not change its constitution or its laws as life and the chromosome arose. This is why I carry the burden of the antithetical nature of matter and this is how the duality of my behaviour was imprinted into my construction.

7

My Split Personality — The Source of "Folly"

This antithetical heritage led me to have a split personality. The result has been that generations of scientists have been confused and bewildered. What kind of creature was the chromosome? Was it plastic, constantly changing by innumerable rearrangements and mutations, or was it a rigid structure, that emerged at every cell division and in every new organism, with the same structural organization and basic functions?

These two opposing properties were most difficult to reconcile.

All the evidence available has shown that the chromosome actually works both ways: it is highly plastic and at the same time highly rigid. My plasticity is unbelievable. For millions of years I have gone through all kinds of structural changes. They have been a normal part of my behavior. I may translocate a whole piece of my body into another part or into another chromosome without this event necessarily affecting my survival. A whole large portion of DNA may be inverted, suddenly locating genes at the other extreme of my body. Small and large deletions as well as duplications of my genes are the order of the day. I even invented "crossing over" which leads to the exchange of segments between the chromosomes of the mother and those of the father, when they pair at the division that precedes the formation of sexual cells (Fig. 1.8).

These rearrangements of structure have been, and are, an obligatory part of my behavior. Modifications of gene function may occur as a result of these translocations and inversions, as well as other exchanges, but my unity of structure and function has endured most of them, since I continue to exist in my original form. What is equally unbelievable is that although I have gone through this untold number of material disruptions, I have at the same time maintained an equally impressive rigidity.

This becomes easily recognizable when one compares the chromosomes of protozoa (like an *Amoeba* or a *Paramecium*) with those of a human being. Two microscopes may be set side by side. If in one is placed a cell with human chromosomes and in the other a cell with protozoan chromosomes, no one will be able to tell who is who (Fig. 1.10). Although humans and protozoa have been separated by evolution for 1.5 billions of years, their chromosomes share the same organization and the same basic functions. Even at the molecular level the identity is startling. The chromosome ends have the same DNA base sequences, and the RNA splicing mechanism is the same in both groups (Lima-de-Faria, 2003).

Moreover, a great number of genes have remained unchanged since the dawn of life. Earlier than the protozoa were the bacteria, and these have the same type of ribosomal RNA genes that are found in humans. Other genes highly conserved, are those responsible for the formation of hemoglobin and chlorophyll. The homeobox genes, involved in embryo development, are not only common to invertebrates and vertebrates, including humans, but are also responsible for flower organization in plants. The conservation has been extreme (Scott, 1992).

More dramatic is that the order of the DNA segments within these genes, which is related to the order of the parts of the body that they affect, has also been preserved throughout evolution. The DNA sequences located in the front part of the gene are those that affect the frontal part of the organism's body, while those located in the rear of the gene are responsible for the caudal region of the body (Fig. 3.6). The rigidity could not be more evident (Lawrence, 1992; Alberts *et al.*, 1994).

It is this split personality that is at the basis of my "folly". There is nothing that I do not dare to change or improvise but at the same

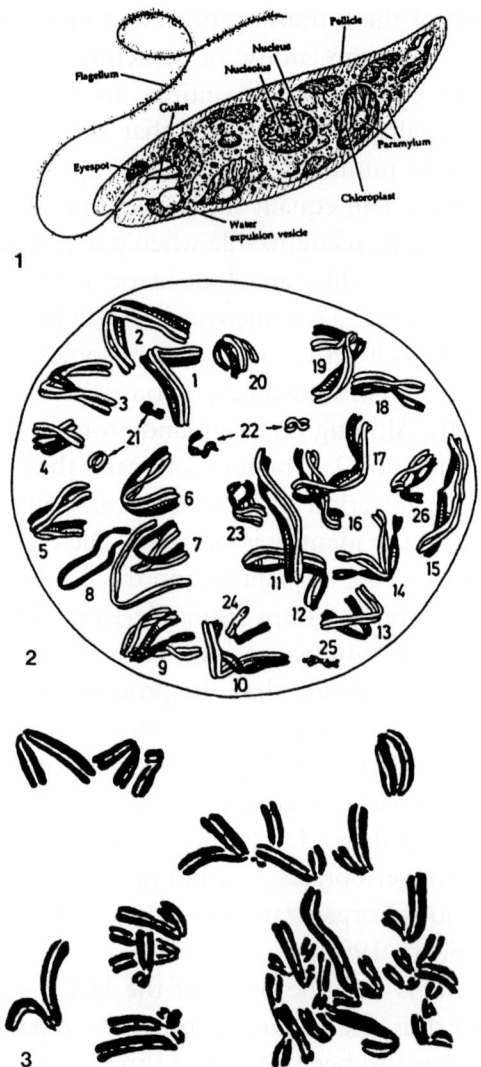

Fig. 1.10 Chromosomes have maintained the same organization and molecular properties since the dawn of the nucleus.

1. The protozoan *Euglena* showing the nucleus, nucleolus and chloroplast.

2. The 26 chromosomes of the protozoan *Barbulanympha ufalula* consist of groups of large, medium size and small chromosomes as are found in the human complement.

3. Human chromosomes from a tumor cell showing variation in number. Their regular number is 46 in most body tissues. The human chromosomes have the same organization, and molecular properties, as those of protozoans.

time I preserve and use constraints. This I do, irrespectively of whether it agrees or not with the outer world that surrounds me. However, in some circumstances I react to the environment, by changing according to its challenges, but this I do mainly when I am being threatened.

Obviously I am a prisoner of my material construction, but within this rigid frame, I have the possibility of going my own way. Who cares for all those conventional ideas that are part of the current lore.

8

The Reason for Confessions

Why should a chromosome write its confessions?

The French author, philosopher and musician Jean Jacques Rousseau (1712–1778) is considered to be the initiator of a new literary genre "The Confessions". Autobiographies had been written before, but the Confessions had another intent. The goal was to reveal the innermost feelings and thoughts as well as to expose both one's weaknesses and strengths.

The proposal to write an autobiography was made to Rousseau by his publisher Marc-Michel Rey in 1761. But he found the idea unappealing.

Rousseau's first reaction was silence. It seemed disgusting to use one's time with such a matter. Was it worthwhile to answer his opponents and enemies who even burned his books? What finally led him to write his confessions was the need he felt to redress some of the insults and accusations that he had suffered throughout his life. He started collecting notes and letters. Later he assembled them, transforming the whole into a book. However, "The Confessions" were first published as a complete edition in 1796, eighteen years after his death.

As Dent (1992) states the book opens with a proud and defiant declaration:

"I have resolved on an enterprise which has no precedent, and which, once complete, will have no imitator. My purpose is to display to my kind a portrait in every way true to nature, and the man I shall portray is myself...I shall come forward with this work in my hand, to present myself before my Sovereign Judge... let the numberless legion of my fellow men gather round me, and hear my confessions. Let them groan at my depravities, and blush for my misdeeds. But let each one of them reveal his heart at the foot of Thy throne with equal sincerity, and may any man who dares, say 'I was a better man than he'."

The reason why the book was not published in his life-time is known. Some parts of the text had been shown to friends who were instrumental in banning the book. They were afraid of being included in it and, in consequence, being exposed in the same way as Rousseau himself. Today the Confessions are treasured as a jewel of the world literature mainly due to the great honesty, sensibility and intelligence that pervade the work.

How dare I the chromosome, a microscopic creature, write my own Confessions especially since I am supposed to have no feelings and no thoughts?

There are two reasons. First, since I am foolish I dare to embark on such an enterprise. I am going to display to my readers a "portrait in every way true to nature". I shall expose my structure and properties and "Let them groan at my depravities, and blush for my misdeeds". I simply cannot do otherwise. But I do not "present myself before any Sovereign Judge" because I am immortal. This is one of the basic properties of my innermost construction. It has allowed me to carry on, as the main bearer of life, for millions of years and I do not intend to stop after such a short journey. But am I really immortal? We shall see how valid is my statement. I am probably boasting.

Second, I have received so many insults and have been the object of a series of false accusations that I cannot remain passive in face of such a public disgrace.

9

I Have Been Abused and Covered with Insults

To be a chromosome is no easy matter. In the last 50 years I have been praised and despised. On one hand, I have been put on a pedestal, being heralded as the sole carrier of the "book of life" and the most important of all cell organelles. On the other hand, geneticists, bewildered for decades with my behavior, resorted to a facile solution considering me: a parasite, a selfish creature, a master of slaves, a nonsense figure and even a junkyard. How can one stand such a collection of insults, without resorting to defence? This army of enemies has been implacable, constantly attacking me and filling pages and whole books with such accusations (Burt and Trivers, 2006).

I am not at the end of my life, but I take this opportunity to redress, not all, but at least some of these mischievous attacks.

When cytologists discovered that besides my regular comrades in the nucleus, there appeared, now and then, some accessory chromosomes in variable number, they wondered what function they could have. Obviously they resorted to the explanation that fitted better in a human society. These extra chromosomes had to be "parasitic". Later, the accessory chromosomes were found to take over the function of sex chromosomes, and to be a reserve of genetic material. Moreover, they were found in large numbers in most species living under natural conditions, a feature that was difficult to dismiss on the grounds of a negative effect on the plant.

Yet the accusation of being parasitic was much more fashionable and it persisted (Camacho, 2004).

When the first measurements of the amount of DNA were performed on chromosomes cytologists became perplexed as they tend to follow naive lines of thought. Since DNA was the genetic material, the more complex organisms should have more DNA, otherwise the increase in the number and size of organs could not be easily explained. A larger brain, with more cells, would need more DNA. Viruses, bacteria and invertebrates, in general, had minor amounts of DNA but when it came to the higher organisms confusion emerged. For example frogs and plants had much more DNA than mammals or humans (Fig. 1.11). This was called the DNA-paradox. Cytologists started thinking that there were two kinds of DNA: a real noble one and another that was just junk. Thus, frogs had to contain a lot of "degenerated" or "junk DNA" (Ohno, 1972).

Chromosomes at certain early stages of division, form large loops that protrude from the main body giving them the appearance of a "lampbrush" (a name that they received when they were discovered and described in detail). These loops were found to be of various sizes, some large others small, and synthesized RNA in a sequence along the chromosome. The disregard for human values resurfaced taking the form of the best explanation. One was dealing with a "Master-Slave" situation. Some regions of the chromosome were the masters that decided the fate of the slaves (Callan, 1967). This time, they were accusing me of no less than pure slavery.

But the situation got even worse. As the methods of DNA analysis became refined, it was discovered that only 3% of the total DNA of the human chromosome complement is used to encode genes. Actually, all the exons in the human genome which are the DNA sequences coding for the proteins, make up only 1.5% of the total (Brown, 2007).

The astonishing amount of 97% is not involved in the formation of structural genes. A shower of insults fell then on me. All this extra DNA had to be not only junk but even "nonsense" DNA.

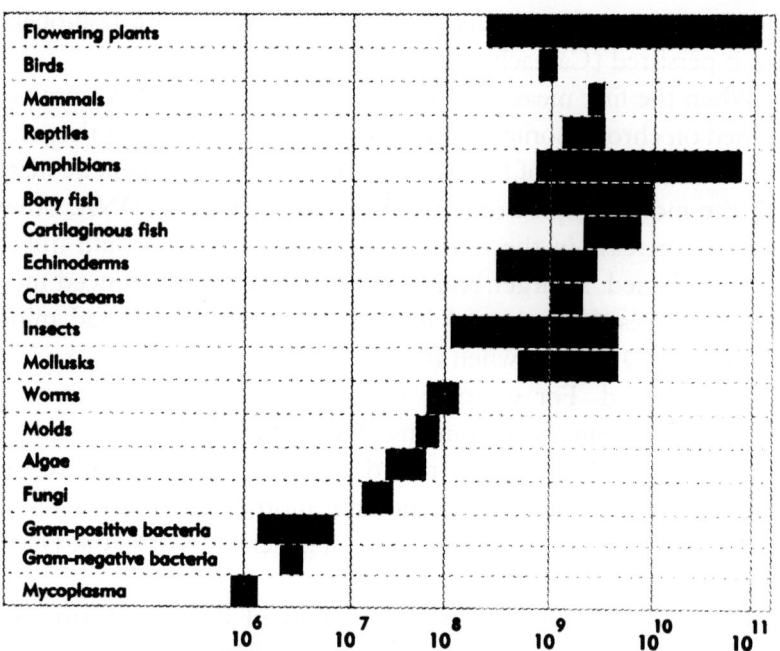

Fig. 1.11 In higher organisms the DNA amount is not related to morphological complexity.

The DNA content of the chromosome complement is related to the structural complexity of lower organisms. It tends to increase between Mycoplasma and the Mollusks. However it varies extensively among other animals and plants. The large variation within a group is in agreement with the molecular behaviour observed in DNA. Gene sequences may have remained unchanged for billions of years, whereas others become modified within days. DNA, like other molecules, contains the antithetic properties of great stability allied to rapid variation. The DNA content mirrors well this dual character of DNA evolution (The genome size is represented in base pairs).

Moreover, a whole book was written accusing me of consisting of "selfish genes" (Dawkins, 1976).

The list does not finish here. I have not only been called "the ultimate parasite" (Orgel and Crick, 1980) but "ultraselfish" by Crow (1988).

10

The Wisdom of the Foolish

Folly contains some of the benefits of insanity. A state of the mind which is usually perceived as having negative effects may turn out to be most valuable under other circumstances.

Every society has involved a certain degree of oppression but the worst tyranny came from the ideas and the values that it imposed on its members. The ruling intellectual elite has had the upper hand and usually did not allow the expression of any idea which was opposed to its tenets. The price has been, most of the time, so high that history books tell us how those who dared to do so payed for such a challenge with their lives. They were either condemned to death, as Socrates (469–399 B.C.) suffered in democratic Athens, or burned at the stake, as was the philosopher Giordano Bruno (1548–1600) for the safeguard of the Christian doctrine. Albert Einstein (1879–1955) had to immigrate to three different countries, and had to acquire different citizenships, to carry out his work under less harassment. The examples could be multiplied.

Others who did not dare to face a full confrontation resorted to satire. This was also a dangerous posture, but it was an indirect threat to the establishment and as such has been throughout the ages, better tolerated, even though it also contained risks. In every period there were those minds that satirized the prevailing mores and dominating ideas of the time.

It is always a pleasure to feel in good company.

Aristophanes (450–386 B.C.) is considered one of the greatest poets of ancient Greek comedy. There are nearly a thousand fragments and citations from his work, but eleven plays have survived nearly intact. In them he attacked magistrates, but his favourite targets were men prominent in politics, poets, musicians, scientists and philosophers. Few were left outside his grip. He even ridiculed Socrates which Plato found inappropriate.

From the Greek, one moved into the Roman society. There was also a leading author Horace (65–8 B.C.) who became the emperor Augustus' poet laureate. The four books of the "Odes" are poems of the highest quality, full of praise for nature and the fine values of life. But he wrote also the "Satires" which were essentially ironic and as a consequence had a strong influence not only during Roman times, but throughout the Middle Ages and the 17th and 18th centuries. It was mainly for his Satires that the Italian poet Dante (1265–1321) admired Horace. They were later imitated by the English poet Alexander Pope (1733).

Erasmus of Rotterdam (1469–1536), who became one of the leaders of the innovative Renaissance thinking, was caught in the early stages of the religious conflict, between Catholics and Protestants, that later led to the bloody 30 years war (1618–1648). The medieval scholastic ideas were being replaced by a rational and direct approach to natural phenomena. Erasmus chose satire. The "Praise of Folly" was published in 1511. In this book he "starts by criticizing everything her creator held dear, and celebrating youth, pleasure, drunkenness and the dizzying sexual desires that created us all". Later sections examined human pretensions and he mocked theologians. Erasmus in his "Folly" dared to praise the simple piety which he saw far removed from the teachings of both organized Churches. Above all he repudiated the sterility of the dominating scholastic thinking. The book led to much controversy.

A parallel movement took place in the arts. During the early Middle Ages, Roman theatres had been closed and destroyed. As the Renaissance emerged in the 1500s, theatre plays were again allowed, but had to assume a form of veiled social satire. This was the "Commedia dell'Arte" which included such classical pantomime figures as: Harlequin, Colombine, Pierrot and others. The players did not spare any one in their ridicule. The "Commedia" became the source of modern opera and theatre (Fig 1.12).

René Descartes (1596–1650), is considered to be, after Plato, the philosopher who made the strongest impact on the European mind. His best known work is the "Discourse of the Method" in which he exposed the principles of rational approach to science. He stated: "It is now some years since I detected how many were the false beliefs that I had from my earliest youth admitted as true, and how doubtful was everything I had since constructed on this basis; and from that time I was convinced that I must once for all seriously undertake to rid myself of all the opinions which I had formerly accepted, and commence to build anew from the foundation,

Fig. 1.12   Some of the leading characters of Commedia Dell'arte.

Troop of comedians which includes (from left) Pierrot, Harlequin and Scapin. Etching by L. Surugue (1719) after a painting by the French artist A. Watteau (1684–1721).

if I wanted to establish any firm and permanent structure in the sciences. I shall at last seriously and freely address myself to the general upheaval of all my former opinions." The subject is equally actual in our days.

During the expansion of colonialism and the growth of the industrial revolution, the period of Enlightenment was led by several brilliant writers. François de Voltaire (1694–1778) was the French philosopher who perceived the need to dispose definitely with the intellectual confusion that still prevailed. Among his central works was "Candide" (1759). This was a crushing satire of the thesis maintained by the German philosopher and mathematician Gottfried Leibnitz (1646–1716) that this world was the best of all worlds.

In our present society we have been dealing with similar types of oppression, in which dominating ideas and values are used to direct our social behaviour. Due to the technological development, the irony has taken other forms. Charles Chaplin (1889–1977), the English movie director, did not hesitate to produce films containing a violent satire of our society. The film "Modern Times" (1936) exposed the lack of respect for the individual factory worker. In the middle of World War II the "Dictator" (1940) unmasked the methods of Nazi policies. As the war finished he made a new film "Monsieur Verdoux" (1947) in which he showed the relationship between war and big finance. His courage costed him dear. Soon he was obliged to emigrate to Switzerland.

More recently Dario Fo (born 1926), the Italian writer and dramatist, has given new actuality to the figures of the "Commedia dell'Arte" using them to satirize our society in his plays: "The Great Pantomime" (1968) and "Mistero Buffo" (1969). He was awarded the Nobel Prize in Literature in 1997. This may be taken to mean that the Nobel Committee had become possessed by the "wisdom of the foolish".

According to most dictionaries a jester was a professional fool employed by a ruler, or king, in the Middle Ages to amuse him with antics, tricks and jokes. The Swedish Academy in its press release justified the prize with the following words: "Dario Fo emulates the jesters of the Middle Ages in scourging authority and upholding the dignity of the downtrodden. For many years Fo has been performed all over the world, perhaps more than any other contemporary dramatist, and his influence has been considerable. He if anyone merits the epithet of jester in the true meaning of that word. With a blend of laughter and gravity he opens our eyes to abuses and injustices in society and also the wider historical perspective in which they can be placed. Fo is an extremely serious satirist with a multifaceted oeuvre. His independence and clear-sightedness have led him to take great risks, whose consequences he has been made to feel while at the same time experiencing enormous response from widely differing quarters."

As the American writer R.W. Emerson (1803–1882) wrote in his Essays: "No man is quite sane: each has a vein of folly in his composition".

11

Scientific Concepts are Prone to Change Throughout Time — The Nature of Science Demands that Previous Ideas be Superseded, as New Technologies Allow a Deeper Insight into Matter

Science is not based on dogmas, truths or facts, as sometimes stated.

Firstly, dogmas are the domain of religions and as such are not to be replaced or discussed. Secondly, science does not either seek the truth, its aim is more modest, it investigates the mechanisms directing nature's transformations. Truth deals with the degree of honesty involved in the description of a given situation. It is part of the scientific method but it is not an aim. Thirdly, facts are events that actually happen but they are the product of an interpretation. The type of interpretation is dependent on the experiment chosen, the degree of the advancement of science, the personality of the scientist and other factors.

The goal of science is the establishment of well defined relationships between phenomena leading to the prediction of novel situations. The real test of scientific endeavour is the demonstration that a certain event actually takes place at a given time and under specified circumstances. Prediction is the corner stone of scientific validity. In chemistry the products of a reaction can be foreseen. In astronomy the exact position of a planet, on a given day, hour and minute, can also be foreseen. In a cell, the division of the chromosomes and the consequent formation of specific types of cells can also be determined in advance.

Science is based on evidence but what is evidence? It is the collection of observations and experiments, which, when assembled in a logic and coherent way, allow the formulation of working hypotheses. These in turn lead to new working hypotheses that disprove or prove the assumptions made. When there is general agreement between the various types of data, and predictions turn out to be valid, a theory is then formulated that embraces most phenomena in a given area of knowledge.

The observations and the experiments must be made with the utmost accuracy. But here lies a big pitfall. The degree of accuracy is totally dependent on the technology available at a given time. The use of the naked eye allows to describe a plant, or an animal, with eye accuracy. But if one uses a magnifying glass, the cells start to be visible and a new domain of research appears. If going one step further, a microscope becomes available, then the realm of chromosomes and other cell organelles becomes suddenly patent. In a further stage, the access to an electron microscope will allow us to see the actual molecules of DNA that build the chromosome. By using one of the latest models of the powerful electron microscopes one may obtain photographs of the single atoms that are part of this macromolecule. Thus, the accuracy and the validity of the observations, are highly dependent on the type of instrument used in the analysis of the object studied. The observations made at the first level with the naked eye were not erroneous, nor were those made with the magnifying glass or the light microscope, and as such, remain valid. It was only the type of information that they revealed that was limited.

This is one of the main reasons why scientific concepts change all the time. They do so, not because there was a fault in the initial approach, but because the observations and experiments are permanently being superseded by novel ones made at new levels of the

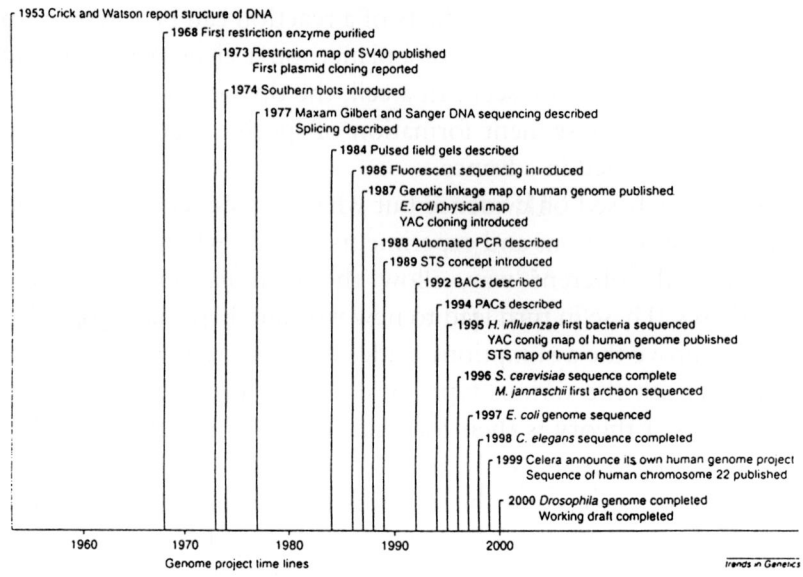

Abbreviations BACs, bacterial artificial chromosomes; PACs, P1 artificial chromosomes; STS, sequence tag sites; YAC, yeast artificial chromosome.

Fig. 1.13   How the development of new technologies led to the modification of genetic concepts.

Table describing some of the results and techniques, published between 1953 and 2000, that led to the sequencing of the genome of various organisms including humans.

organization of matter that had not been accessible previously and permit a better knowledge of the mechanisms involved (Fig. 1.13).

The edifice of science is under constant reshaping, as hypotheses and theories are discarded to be substituted by new ones. There lies the strength of its endeavor. What is permanent is the exactness of science's methodology combined with the unbending logic used in enquiry. These are not changeable because they form its backbone.

12

What Seems Ludicrous at a Given Time Turns Out to be the Correct Explanation Several Years Later. The Gene was Considered to Consist of Protein, But is Now Known to be a Ribbon of DNA

The disciples of Linné, jumping around in the forests of the Amazonas river, in the 18th century — holding in their hands their butterfly nets — were bewildered by the profusion of insect species. The variety of colours and forms, as well as the astonishing number of new species, left them with a sense of the complexity and immensity of nature. The human body is also highly complex. Anatomical studies revealed a great number of tissue types that formed a compilation of no less than 10^{13} cells (Varmus, 2002). At the same time the physiological data disclosed that its orchestration was due to a large number of molecular signals. Again, complexity was the word of the day.

When the chromosome and the gene arrived they were included in this scenario. They had to be equally complex. The nucleic acids had long been known to be part of chromosomes. But they were considered to be chemically too simple to be the source of such an important entity that would build the marvelous human body.

The gene had to consist of proteins. These were well known to be large macromolecules that occurred in many distinct forms. They had to be the material basis of the gene. Actually they constituted the main bulk of the chromosome.

This is why, during World War II (1944), when Avery and collaborators, demonstrated that DNA was the carrier of genetic information in bacteria, geneticists did not take note of this finding. No one had seen chromosomes in bacteria, they reasoned, and humans could not be compared to such microscopic organisms.

Even after the Watson and Crick model of DNA was published in 1953, most geneticists reacted with scepticism. It was not until the 1960s, following a series of novel experiments, that DNA became accepted as the carrier of genetic information.

13

In the Last 20 Years the Number of Human Genes was Reduced from 200,000 to 32,000, and this Figure Remains Uncertain

What was the number of genes that shaped the human body? It had to be large.

The first evidence from the genetics of bacteria, revealed that a segment of DNA produced a sequence of RNA which in turn gave rise to one protein. The number of proteins known to exist in humans was no less than 200,000. Geneticists concluded that, on the basis one DNA — one protein — the number of genes should be at least 200,000.

Already in the 1980s, it became evident that one single DNA segment could produce several proteins, and as a consequence the number of genes could be much smaller, but few considered the issue. As late as 2000 it was still written that humans had 150,000 genes (Gilbert, 2000). But soon this figure started to dwindle to 100,000 and 70,000. The sequencing of the bases of the human genome gave an unexpected answer: 39,114 genes (Celera) and 31,780 (The Public Sequence) (Bork and Copley, 2001). Since then these values have sunk to 20,000 (Pennisi, 2003).

It is highly probable that this figure is still too large for two reasons.

Firstly, single genes can produce, not only a few, but hundreds of different proteins as is the case of the neurexin genes (Ullrich *et al.*, 1995). Secondly, many of the proteins found in the human body can be grouped into 1,000 families due to their basic similarities (Dayhoff *et al.*, 1975; Chothia, 1992). Hence, the number of primary genes responsible for the large amount of proteins may be as small as 1,000.

14

The Models of Chromosomes Have Varied Drastically with Time

Another way in which science works is by building models and the chromosome has not escaped this procedure.

Models have advantages and disadvantages.

The positive side of a diagram is that it simplifies and concretizes a dispersed body of knowledge that had not before been condensed into a single picture. Another advantage is that by formalizing several concepts it obliges researchers to find out whether the qualifications assigned to the model are actually correct. In this way it promotes further research which is targeted at validating or invalidating the model.

The negative aspect is that a diagram is always a simplification. Moreover, it often includes acquired misconceptions or factual errors which tend to be perpetuated in this way. This makes it dangerous because, time and again, it is reproduced, being accepted as the final interpretation of a given structure or phenomenon.

This is exactly the case with chromosome models. They have been valuable because they led to further research directed at testing their validity, and they had a negative effect because they perpetuated misconceptions.

Koltzoff (1928) was a visionary. He synthesized the little knowledge available at the time into a diagram with several novel features. The model included the two chromatids of a chromosome as basic units of replication, a feature that had not been clearly defined before. Also, he represented the chromosome as consisting of a large number of protein molecules which were part of its body and of its cellular environment, emphasizing in this way its molecular activity. The most original feature was that he represented each chromatid as consisting of parallel aligned molecules which built two identical fibers, a prerequisite for chemical copying (Fig. 1.14).

It is striking to note that most chromosome models that appeared in the thirties, forties, and even in the fifties, lacked the molecular component and did not include the chromosome replication.

What is new in these models is that they include the spiral structure of the chromosome which had been so well depicted and photographed by many different cytologists, both in plants and animals. Also by this time, three regions of the chromosome had been found to be of general occurrence. These were the centromere, the nucleolus organizer and the telomeres.

These models also taught us how entrenched ideas, that were not well established, took time to die out. Heitz's model of 1935 includes the matrix as a main component of the chromosome body. But its existence was never established, being abandoned in the 1960s.

The chromosome pellicle was considered a necessary structure that enveloped the chromosome body delimiting it well from the nuclear sap and from the cytoplasm. The pellicle is included in the model of Schrader (1944). This thin sheet was actually never observed and soon it became evident that the chromosome had no limiting membrane.

One of the first models to include the different levels of organization from the gene to the metaphase chromosome was that of Darlington and Mather (1949). The nucleotides were included in the scheme for the first time, but the gene was located not as a part of their chemical composition, but was drawn at the side, anchored on the protein fiber. Hence the gene continued to consist of protein. But this model represented a big step forward, for it not only described all the organizational levels, but it also included the dimensions of the various chemical and structural components (Fig. 1.14).

An uncertain feature was the replication of the DNA molecule along the chromosome. It was difficult to conceive that such a molecule could be so long as to be continuous throughout the whole chromosome length. Even after the use of tritiated thymidine in the study of DNA replication, models were devised, in which protein linkers were included between DNA molecules. But by the middle 1970s this concept was dead. Petes (1974), and others, had measured the length of the DNA fiber in single chromosomes and demonstrated the continuity of the DNA molecule along the whole chromosome body (Kavenoff and Zimm, 1973).

It also became evident that the DNA molecule crossed the centromere from one arm to the other, but the division cycle of this region remained a source of confusion. Cytologists, who obtained good preparations of chromosomes, could see that the centromere, which was attached to the spindle, was divided at metaphase and that it was the proximal regions of the arms that held the chromatids together. However, many recent textbooks still present models with the centromere undivided at metaphase, an archaic notion based on the defective cytology of the 1930s.

A structural feature that had been discarded since the 1940s was the chromosome scaffold which consisted of nonhistone proteins. These proteins had been forgotten and were not included in any models. It was the electron microscopy work of Paulson and Laemmli (1977) that showed that the scaffolding proteins were the molecular component that maintained the chromosome's framework. As a result the scaffold proteins became incorporated in the new models.

But surprisingly there is an impressive amount of information on chromosome properties which has been left out of even the most recent graphic representations. This is apparently due to the

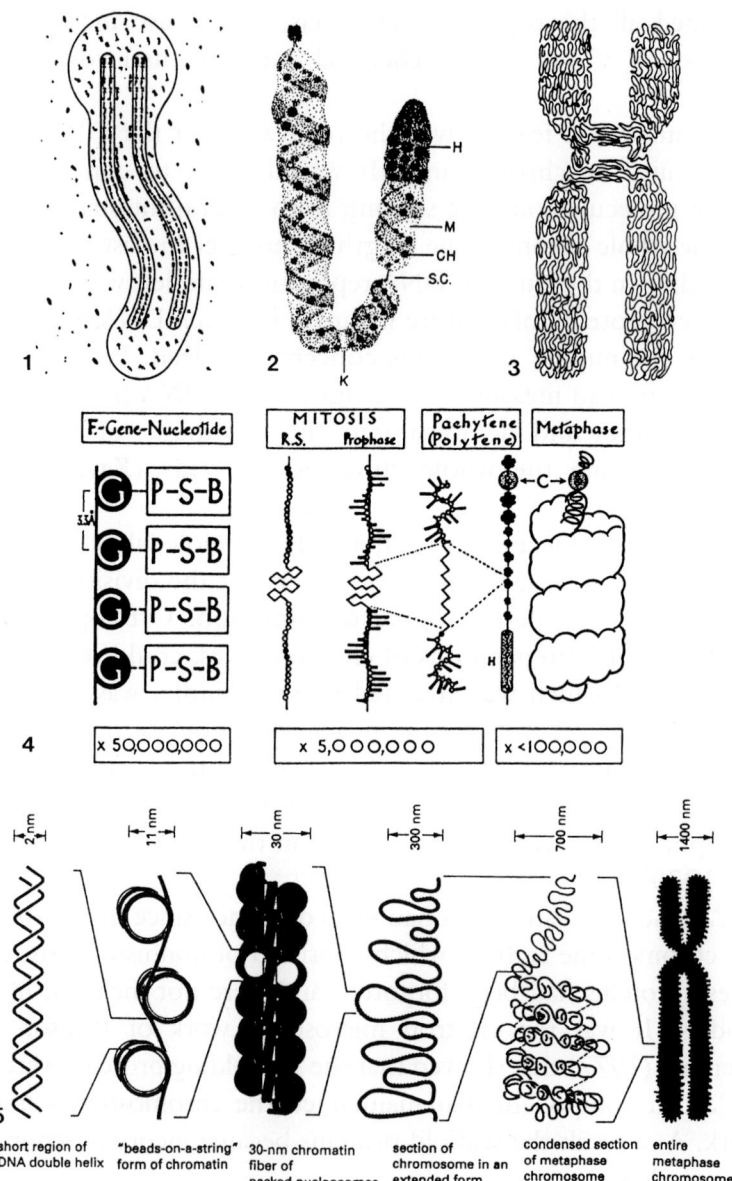

Fig. 1.14   How chromosome models have changed, and continue to do so, as new information accumulates.

1. Koltzoff 1928 — The first model of a chromosome depicting its chemical composition as well as its separation into two chromatids as a consequence of a biochemical process. Each

absence of a coherent picture that integrates the interactions between the different regions. The infomation that is mainly missing has to do with the function of the chromosome as a whole. Without it, it seems difficult, if not impossible, to discover the mechanism responsible for the maintenance of chromosome integrity.

---

chromatid was considered to be composed of aligned molecules building two identical fibers. In 1928 the genetic material was supposed to consist of proteins. Nucleic acids were thought to be too simple to harbour the complex structure of the gene.

2. Heitz 1935 — Model of the chromosome including the main cytological features of its structure. CH, chromomere; H, heterochromatin; K, kinetochore or centromere; SC, secondary constriction (sometimes identical to nucleolus organizer); M, matrix. Until the 1960s the existence of a matrix was under debate.

3. Comings 1972 — Single-stranded model of chromosome structure. A single DNA-protein fiber is considered to begin at one telomere, to fold upon itself, to build up the width of the chromatid and eventually to progress to the opposite telomere. The centromere is represented well divided in the metaphase chromosome. The chromosome consists of two chromatids, united on both sides of the centromere, which will separate to opposite poles.

4. Darlington and Mather 1949 — The size relations and supposed systems of arrangement, molecular and super-molecular, of gene-elements (G), nucleotides (P-S-B), chromosomes and chromosome spirals at mitosis and meiosis. F, protein fiber; RS, resting stage; H, heterochromatin; C, centromere; PSB, phosphate-sugar-base. This is the first scheme to cover the organization of the chromosome from the molecular level of the gene to the supercoiled metaphase chromosome, showing the main intermediate stages. Note that the gene is attached to a protein fiber and separated from the nucleotides. In 1949 the gene of higher organisms continued to be considered to consist of protein and not of DNA.

5. Alberts *et al.* 1994 — Model of chromatin packing showing some of the successive orders of organization postulated to give rise to the highly condensed metaphase chromosome. The DNA becomes packed with histones in the form of nucleosomes which associate building successive levels of spirals. In this model nonhistone (or scaffolding) proteins are not represented.

References

The references follow the order in which they appear in the text. But when a reference has been included before it is not repeated.

Part I

Waldeyer W (1888) Über Karyokinese und ihre Beziehungen zu den Befruchtungsvorgängen. *Archiv für mikroskopische Anatomie und Entwicklungsmechanik* **32**: 1–122.

Page SL, Hawley RS (2003) Chromosome choreography: the meiotic ballet. *Science* **301**: 785–789.

Luria SE *et al.* (1978) *General Virology*, 3rd ed. Wiley, USA.

Bajer A (1957) Cine-micrographic studies on mitosis in endosperm. III. The origin of the mitotic spindle. *Exp Cell Res* **13**: 493–502.

Carrington JC, Ambros V (2003) Role of microRNAs in plant and animal development. *Science* **301**: 336–338.

Mello CC, Conte Jr, D (2004) Revealing the world of RNA interference. *Nature* **431**: 338–342.

Wood RD (1996) DNA repair in eukaryotes. *Annu Rev Biochem* **65**: 135–167.

Burrows A (2000) Supernova explosions in the universe. *Nature* **403**: 727–733.

Hillebrandt W *et al.* (2006) How to blow up a star. *Sci Am* October 2006: 23–29.

Alfven H (1966) *Worlds — Antiworlds. Antimatter in Cosmology.* Freeman, San Francisco, USA.

Pagels HR (1982) *The Cosmic Code. Quantum Physics as the Language of Nature.* Michael Joseph, London.

Quinn HR, Witherell MS (1998) The asymmetry between matter and antimatter. *Sci Am* October 1998: 50–55.

Woosley S, Janka T (2005) The physics of core-collapse supernovae. *Nature Physics* **1**: 147–154.

Lima-de-Faria A (2003) *One Hundred Years of Chromosome Research and what Remains to be Learned.* Kluwer Academic Publishers (2003) Boston, USA and Springer (2004) Berlin, Germany.

Scott MP (1992) Vertebrate homeobox gene nomenclature. *Cell* **71**: 551–553.

Lawrence PA (1992) *The Making of a Fly.* Oxford Blackwell Sc. Publ., London, UK.

Alberts B *et al.* (1994) *Molecular Biology of the Cell.* Garland Publishing Inc., New York, USA.

Dent NJH (1992) *A Rousseau Dictionary.* Blackwell, Reference, London, UK.

Burt A, Trivers R (2006) *Genes in Conflict. The Biology of Selfish Genetic Elements.* Harvard Univ. Press, Cambridge, MA, USA.

Camacho JPM (editor). (2004) B Chromosomes in the eukaryotic genome. *Cytogen Genome Res* **106** (Nos. 2–4). Karger, Switzerland.

Ohno S (1972) So much "Junk" DNA in our genome. In: Smith HH (ed.), *Evolution of Genetic Systems.* Gordon and Breach, NY, USA, pp. 366–370.

Callan HG (1967) On the organization of genetic units in chromosomes. *J Cell Sci* **2**: 1–7.

Brown TA (2007) *Genomes 3.* Garland Science, NY, USA.

Dawkins R (1976) *The Selfish Gene.* Oxford Univ. Press, Oxford, UK.

Orgel LE, Crick FCH (1980) Selfish DNA: the ultimate parasite. *Nature* **284**: 604–607.

Crow JF (1988) The ultraselfish gene. *Genetics* **118**: 389–391.

Erasmus of Rotterdam, 1511. *Praise of Folly.* Penguin Books (1993), London, U.K.

Descartes R (1997) *Discourse of the Method.* Key Philosophical Writings. Wordsworth Classics, UK.

The Nobel Prize for Literature 1997 — Dario Fo. Swedish Academy. The Permanent Secretary. Press Release, October 9, 1997.

Emerson RW (1841 and 1844) *Essays.* First and Second Series. Oxford University Press (1950), London, U.K.

Varmus H (2002) What does knowing about genomes mean for science and society. In: Yudell M and DeSalle R (eds.), *The Genomic Revolution.* Joseph Henry Press, Washington, DC, USA, pp. 20–34.

Avery OT *et al.* (1944) Studies on the chemical nature of the substance inducing transformation in pneumococcal types. *J Exp Med* **79**: 137–158.

Watson JD, Crick FHC (1953) Genetical implications of the structure of deoxyribonucleic acid. *Nature* **171**: 964–967.

Gilbert SF (2000) *Developmental Biology*, 6th ed. Sinauer Associates Sunderland, MA, USA.

Bork P, Copley R (2001) Filling in the gaps. *Nature* **409**: 818–820.

Pennisi E (2003) Gene counters struggle to get the right answer. *Science* **301**: 1040–1041.

Ullrich BYA *et al.* (1995) Cartography of neurexins: more than 1000 isoforms generated by alternative splicing and expressed in distinct subsets of neurons. *Neuron* **14**: 497–507.

Dayhoff MO *et al.* (1975) Evolution of sequences within protein superfamilies. *Naturwissenschaften* **62**: 154–161.

Chothia C (1992) One thousand families for the molecular biologist. *Nature* **357**: 543–544.

Koltzoff NK (1928) In 1939. Les Molecules Heréditaires II. Actualités Scientifiques et Industrielles No. 776: 48. Hermann and Co., Paris, France.

Heitz E (1935) Chromosomenstruktur und Gene. *Z. Indukt. Abst. Vererbungsl.* **70**: 402–447.

Schrader F (1944) *Mitosis.* Columbia University Press, NY, USA.

Darlington CD, Mather K (1949) *The Elements of Genetics.* George Allen and Unwin, London, UK.

Petes TD *et al.* (1974) Yeast chromosomal DNA: size, structure and replication. *Cold Spring Harbor Symp. Quant Biol* **38**: 9–16.

Kavenoff R, Zimm BH (1973) Chromosome-sized molecules from Drosophila. *Chromosoma* **41**: 1–27.

Paulson JR, Laemmli UK (1977) The structure of histone-depleted metaphase chromosomes. *Cell* **12**: 817–828.

Sources of Illustrations

Part I

1.1 Lehninger AL (1975) *Biochemistry*, 2nd ed. Worth Publishers Inc., New York, USA. (Fig. 1.1 page 19).

1.2 (1) Selective painting of human chromosomes. Image by courtesy of H. Scherthan, Munich, FRG.

    (2) Mottier DM (1903) *Bot Gaz* **35**: 250–280. (Fig. Cell division of *Lilium*).

    (3) DuPraw EJ (1970) *DNA and Chromosomes*. Holt, Rinehart and Wilson, New York, USA. (Fig. 1 page 72).

1.3 Read H (1960) *Det Moderna Måleriets Historia* (Swedish translation). Albert Bonniers Förlag, Stockholm, Sweden. (Plate on page 35).

1.4 Néret G (2006) Henri Matisse. Taschen, Köln, London. (Fig. "Woman before an Aquarium" page 113. The Art Institute of Chicago, Chicago (IL.), USA).

1.5 (1) Bridges CB (1935) *J Heredity* **26**: 60. (Fig. The chromosome 4 of *D. melanogaster*).

    (2) Rückert J (1891) *Anatomischer Anzeiger*, Vol. 1891. Jena, Germany. (Fig. Chromosomes of the germinal vesicle).

1.6 Lima-de-Faria A *et al.* (1959) *Hereditas* **45**: 467–480. (Figs. 26-40, page 475).

1.7 Courtesy GE Palade (1970) In: Loewy AG and Siekevitz P, *Cell Structure and Function*. Holt, Rinehart and Winston, London, UK. (Fig. 4–28, page 39).

51

1.8 (1) Darlington CD (1937) *Recent Advances in Cytology.* Churchill J and A, London, UK. (Fig. 21A, page 88).

(2) Lima-de-Faria A (1952) *Chromosoma* **5**: 1–68. (Fig. 23, page 24).

1.9 (1) Alfven H (1966) *Worlds-Antiworlds, Antimatter in Cosmology.* Freeman, San Francisco, USA. (Fig. 5, page 26).

(2) Quinn HR and Whitherell MS (1998) *Sci Am,* October 1998 (Fig. Composite particles, page 52).

1.10 (1) Brusca RC and Brusca GJ (1990) *Invertebrates.* Sinauer Associates, Sunderland, USA. (Fig. 4A, page 132).

(2) Cleveland LR (1949) *Trans Amer Phil Soc* **39**:1 (Fig. Prophase chromosomes).

(3) Heim S and Mitelman F (1987) Cancer Cytogenetics. Alan R. Liss., New York, USA. (Fig. 1, page 4).

1.11 Lewin B (1994) *Genes V.* Oxford University Press, Oxford, UK. (Fig. 22.1, page 658).

1.12 Oreglia G (2002) Commedia Dell'Arte. Ordfront, Stockholm, Sweden. (Fig. Komedianttrupp, page 17).

1.13 Dunham I (2000) *Trends in Genetics* **16**(10): 458. (Fig. 1, page 458).

1.14 (1) Koltzoff NK (1928) In: 1939, *Les Molecules Hereditaires II. Actualités Scientifiques et Industrielles* No. 776. Hermann et Co. Editeurs, Paris, France. (Fig. 19, Planch III, page 48).

(2) Heitz E (1935) *Z Indukt Abst U Vererbungsl* **70**: 402–447. (Fig. 10, page 425).

(3) Comings DE (1972) *Adv Human Gen* **3** (Fig. chromosome model, page 237).

(4) Darlington CD and Mather K (1949) *The Elements of Genetics.* George Allen and Unwin, London, UK. (Fig. 36, page 148).

(5) Alberts B *et al.* (1994) *Molecular Biology of the Cell.* Garland Publ, New York, USA. (Fig. 8–30, page 354).

Who Cares for Gravity

15

The Chromosome in Its Organization and Activity Follows Its Own Path — It Does Not Obey Gravity, Randomness, Selection or Magnetism

We are so accustomed to take for granted, that all material bodies tend to fall to the ground, that no one dares to think that anything else can happen that deviates from this behavior.

The chromosome is as much a material body as any other structure. It is a huge molecular edifice consisting of millions of atoms. The weight of its nucleic acids and proteins, that are built by these atoms, can be measured accurately. Hence, a chromosome should always fall to the ground, but it does not.

The same is true for randomness. In its permanent exchange of atoms and molecules, the chromosome would be prone to be a prey to permanent random events. Yet the chromosome has created, not only one, but a series of mechanisms which dispose with randomness, otherwise it would not have existed today.

Also, the same holds true for selection. This abstract system of choice may be of interest in social events, but it is not significant for a highly coherent chromosome which has developed mechanisms that easily circumvent selection. The organism or the cell may die, but the chromosome manages to survive and to maintain its coherence using its own private solutions.

Magnets have a polarity that is not disrupted when they are divided into minor units or when the small units are reassembled into larger ones. Chromosomes are not magnets, yet they share some of their properties. They have their own polarity and besides are capable of disassembling into minor units and reassembling into larger ones, which remain fully functional. This they achieve by innovation, creating new terminal regions and readjusting their gene activities.

The chromosome follows its own pass ignoring the interpretations of its behavior based on the traditional explanations.

16

Definition of Gravity — Newton's Laws are Good for Planets and Apples

In physics gravity is: "The force that tends to draw all bodies toward the center of the earth". Gravitation covers a more general situation. It is: "the force by which every particle of matter tends to approach every other particle in the universe" (Webster, 1976).

Physics became an experimental science when the motion of bodies was studied in detail by Galileo Galilei (1564–1642). At the same time astronomy acquired a precise status when Johann Kepler (1571–1630) described the ordered movement of planets. Later Isaac Newton (1642–1727) combined their information into the laws of motion and gravitation that he enunciated in 1687. "The gravitation force that is exerted on the sun and the planets is directly related to their total mass. This force acts at long distances and it diminishes with the inverse of the square of the distance".

According to his beautiful young niece Catherine Barton, who was his housekeeper, the idea of the Moon's gravitation around the Earth stemmed from seeing an apple fall from a tree beside his Woolsthorpe home where he was born. His law could be applied to planets but to apples as well. Newton's theory, had the great value of unifying the movement of the celestial bodies with that of matter on earth. Due to its mathematical formulation and simplicity it represented an impressive triumph of the discovery of order in the universal construction (David *et al.*, 2002). Among other things it explained the Earth's tides. In the 1900s it allowed the determination of the masses of molecules and atoms, and lately it led to the calculation of the accurate trajectories of space vehicles. The laws led to predictions, that were soon verified, but at the same time exceptions were discovered.

Newton, himself, was not satisfied with his theory because it implied an interaction at distance by a "force" that could not be isolated. He allowed for the possibility of errors in measurement and deviations from perfect experimental conditions, but he felt at heart, like later Einstein, that Nature systematically follows exact laws.

The exceptions turned up soon. The planets Neptune and Pluto were discovered from perturbations in the movements of their neighbors. Newton's law turned out to be slightly incorrect near the sun, where the effects of the solar mass modified the orbit of Mercury (Dicke, 1981).

Newton's theory also failed, and was replaced by Einstein's general relativity, when the extreme distances of galaxies were analyzed. This result led Einstein to the substitution of gravity by a curved space concept. Gravity would arise from local conditions of space (Hey and Walters, 2003).

As the American physicist Hans C. von Baeyer (1992) stated "Today physicists regard Newtonian action at distance as obsolete". Gravity, like electricity, is considered to be transmitted by particles that travel through matter and empty space.

17

What is a Force — The Four Fundamental Forces

Force is defined "as any action that alters, or tends to alter, a body's state of rest or of uniform motion" (Pitt, 1988).

The term force is mainly an expression of ignorance. It only represents the recognition of a phenomenon whose nature and mechanism is ignored or not immediately evident. It has been, however, most valuable, because it represented a first step into the enquiry of a phenomenon.

At present four fundamental forces are considered sufficient to explain physical events: gravity, the "weak" force, electromagnetism, and the "strong" force. Gravity holds the stars together and obliges objects to fall to the ground. The weak force is responsible for the release of particles in radioactivity. The electromagnetic force binds electrons into atoms and drives chemical reactions. The strong force glues neutrons and protons together into atomic nuclei. Gravity is the most elusive of the four because it pervades all matter (Glashow, 1997) (Fig. 2.1).

Most important for our understanding of the particular behavior of the chromosome, is that the gravitational forces were found to be overshadowed by the strong and weak forces of the atomic nucleus. It is actually electromagnetism, together with these two forces, which have the dominating role in deciding the behavior of atoms and molecules outside and inside the cell.

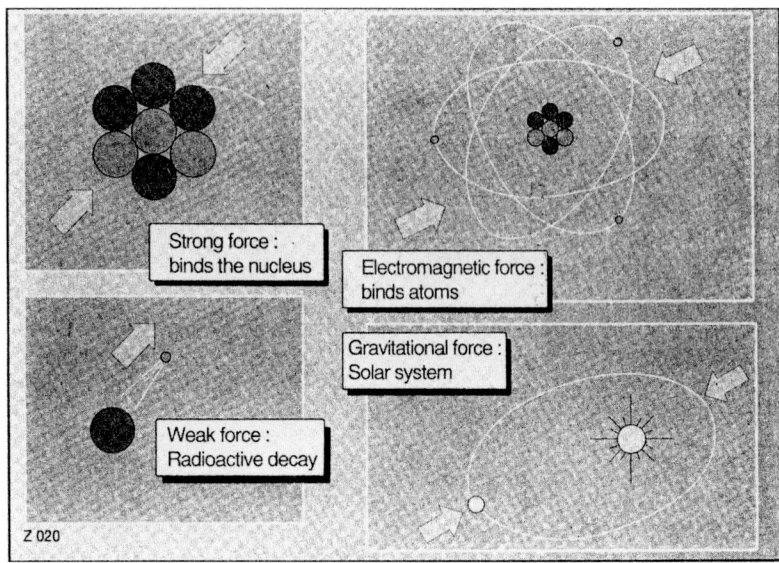

Fig. 2.1   The four forces that physicists consider responsible for matter interactions. The strong force binds the particles that constitute the atomic nucleus such as protons and neutrons. The weakforce is responsible for radioactivity allowing the escape of particles and radiation from atoms. The electromagnetic force locates electrons in orbitals around the atomic nucleus and leads to the binding between atoms. The gravitational force is responsible for the interactions among the components of the solar system.

Of the four forces, gravity is the weakest. As Pagels (1982) put it: "Einstein saw that gravity was a superfluous concept — there isn't any "gravitational force". What actually happens is that the mass of a planet — or any mass — curves the space near it, altering its geometry". For Einstein gravity was not a force but a geometric event (Clark, 1973).

18

Newton's Laws Do Not Apply in Quantum Mechanics

The period 1905–2005 saw an explosion in the development of physics that led to the discovery of the internal organization of the atom and of a plethora of elementary particles that have so far been classified into two groups: quarks and leptons. Their behavior has been difficult to define with precision. Moreover, in the microscopic domain it turned out that a new mechanism different from classical mechanics had to be considered.

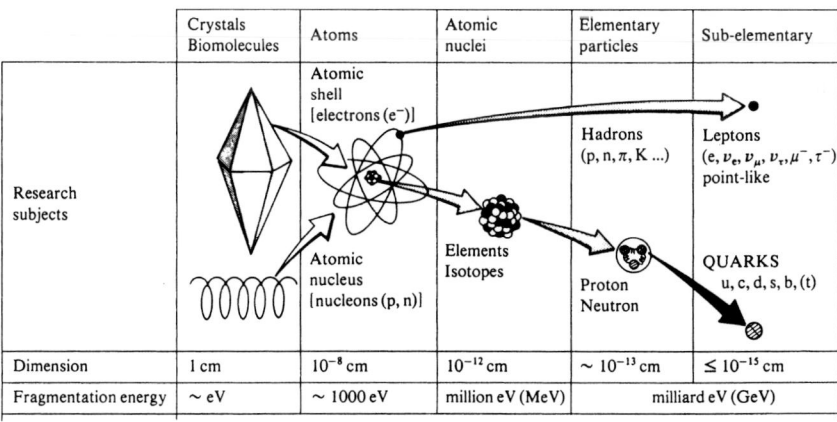

Fig. 2.2   Successive stages in the discovery of the nature of matter.

The progression of research from crystals and biomolecules to sub-elementary particles, showing their dimensions and fragmentation energies.

As Hey and Walters (2003) point out "Newton's laws of classical mechanics must give way to quantum theory" when it comes to elementary particles. Even more, they also add that "There is another area where Newton's laws have been shown to be in need of modification, namely, when the velocities of objects are close to the speed of light". Hogan (2007) puts it plainly: "Gravity feels like a force you can trust. Every day, unwavering, it keeps your feet planted firmly on the ground. But for many physicists, gravity is unsettling. It's the one force that doesn't fit into their quantum picture of the world."

Newton's classic laws remain valid for the planets but are not necessarily valid for the minor particles of matter such as photons or bosons (Fig. 2.2).

19

Not All Bodies Fall When Unsupported

The Greek scientist and philosopher Aristotle (384–322 B.C.) with his universal mind pondered over most phenomena found in nature. He already noticed that most bodies had the ability to fall to the earth surface when unsupported. But the quality was not universal, fire moved naturally upwards.

Capillarity occurs when a liquid ascends in tubes of fine bore. This movement has been attributed to cohesion, adhesion and surface tension in liquids which are in intimate contact with solids. Giovanni Borelli (1608–1679) demonstrated that the level reached by the liquid was dependent on the tube's internal diameter. He dismissed an explanation of the ascent based on reduced atmospheric pressure because capillarity occurred in a vacuum. Laplace (1749–1827) proposed that an exact law of attraction was not necessary to explain the phenomenon which today still remains clouded by uncertainty.

The capillarity rule formulated by Juvin only describes the elevation of the liquid in a tube, but does not elucidate the phenomenon in molecular terms (Pitt, 1988).

20

The Ascent of Sap in Trees — Another Unexplained Phenomenon

Trees can be enormous plants with innumerable leaves. The increase in the aerial shoot demands the absorption of large amounts of water by the roots since trees reach impressive heights.

A young redwood tree, that is 45 m long, uses about 600 kg of water each day, a figure that increases appreciably as the plant gets bigger (Dawson, 1988).

Some of the world's tallest trees are known from the forests of the U.S.A. and Canada. The redwood (*Sequoia sempervirens*) with a height of 112.7 m is equivalent to a 30 storey building. The Douglas fir (*Pseudotsuga menziesii*) rises to 100 m and the sitka spruce (*Picea sitchensis*) may measure 95 m. These are, however, surpassed by the Australian species of *Eucalyptus* trees measuring more than 130 m in height (Woodward, 2004).

The significant event is that the sap moves against gravity, along the stem of the plant, to every leaf located at this height. Initially it was considered that capillarity was responsible for this counteraction. But soon it turned out to be a naive explanation, that would never allow water to be delivered in the quantities required.

The next approach was that the force originated in the strong pressure that occurred in the roots. It soon turned out that this would never be sufficient to raise water to the top of a tall tree. As Hopkins and Hüner (2004) point out "Root pressure clearly cannot serve as the mechanism for the ascent of sap in all cases".

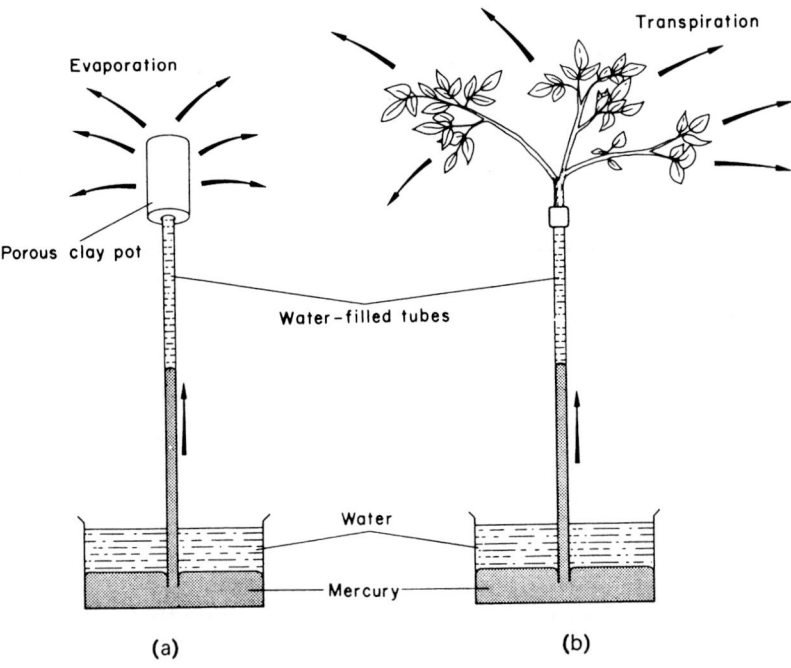

Fig. 2.3   The cohesion-tension interpretation of sap ascent in plants.

The evidence demonstrates the pull up of a column of mercury on a tube, but does not explain the sap movement to the top of trees that are more than 100 meters high. (a) Physical system for the demonstration of the cohesion-tension theory of sap ascent. Evaporation of water from the porous pot provides the tension required to pull up the column of mercury. (b) Transpiration of the leafy twig provides the tension necessary to pull up the column of mercury.

Another alternative was that water could be pulled up by the forces of evaporation occurring in the leaves due to a combination of water cohesion and tension. Even this explanation is not fully accepted because many questions of sap transport remained unanswered (Fig. 2.3).

Radioactive tracers, fluorescent dyes and other devices have been used to check on the various alternative explanations. So far they have failed to give a reliable description of the ascent of sap to 120 meters along the stem of a plant (Richardson, 1975; Bidwell, 1979).

This phenomenon represents an impressive way by which living organisms bypass the universal laws of gravitation.

This problem gets even more intriguing when one analyzes the transfer of the sap over long distances between different organs of the same plant.

An organ, such as a leaf, produces more nutrients than it requires for its own metabolism and growth, due to its photosynthetic power. It thus exports a large part of its assimilate. Roots, stems and fruits, on the other hand are in great need of these nutrients and import them by translocation across the plant vessels. Assimilate transport has been explained by: (1) simple diffusion, (2) cytoplasmic streaming, (3) ion pumps, (4) contractile elements and other proposals. However, all of them "have been largely rejected on both theoretical and experimental grounds" (Hopkins and Hüner, 2004).

Even more confusing is that the translocation of the nutrients occurs simultaneously in opposite directions, a finding that is incompatible with the pressure-flow hypothesis and as such rejects it.

21

An Unknown Process Decides Which Cells are Going to Grow in the Direction of Gravity and Which are Going to Counteract It

Plants furnish an equally astounding case that is seldom or never mentioned in physiology books. The fossil record, and other information on evolution, teaches us that algae were the first plants which emerged around 1,600 million years ago. These algae which initially floated, both in marine and fresh water, soon established a polarity. Some species got fixed to the sea bed by a root-like outgrowth and produced branches that grew upwards.

When plants invaded the land this polarity became sharply transformed into a dichotomy. The consequences were dramatic. About half of the plant followed the action of gravity building what is called the roots. The other half, called the stem followed a direction opposing gravity. This second part not only ignored but actually counteracted this force (Fig. 2.4).

The crucial moment of separation between these two states occurred already in the first cell divisions of the embryo. In maize, a cross section of the embryo shows clearly how it splits into two opposing regions, one that is growing to become the stem and the first leaf (coleoptile), and the other growing in the opposing direction to become the root (radicle) (Fig. 2.5).

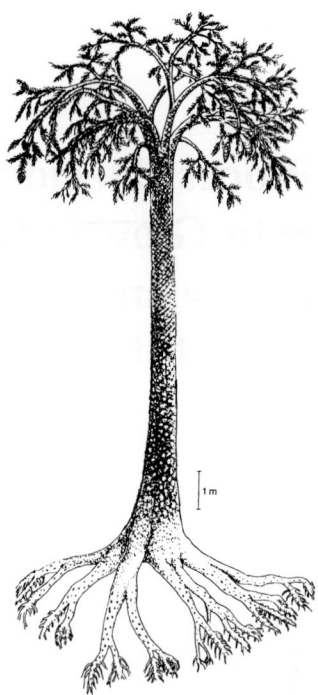

Fig. 2.4   The first trees that appeared on earth already had a clear cut division into gravity and antigravity growth.

The *Lepidodendron* was among the earliest trees that are part of the fossil record (circa 380 million years ago). Its rooting system was as impressive as the above-ground of the plant. It consisted of four, or more, radiating arms that were extensively branched and extended for over 12 m in length below ground.

Plant hormones have been extensively studied and they are known to influence many functions such as plant growth. There are not less than five, but the most important are auxins.

One could imagine that auxins would be responsible for the origin of this dichotomy. However, it turns out that in an oat seedling, this growth hormone has a high concentration both in the actively growing tip of the coleoptile and in the root apex. Hence, the auxin distribution does not throw light on the mechanism behind this difference in the fate of the cells.

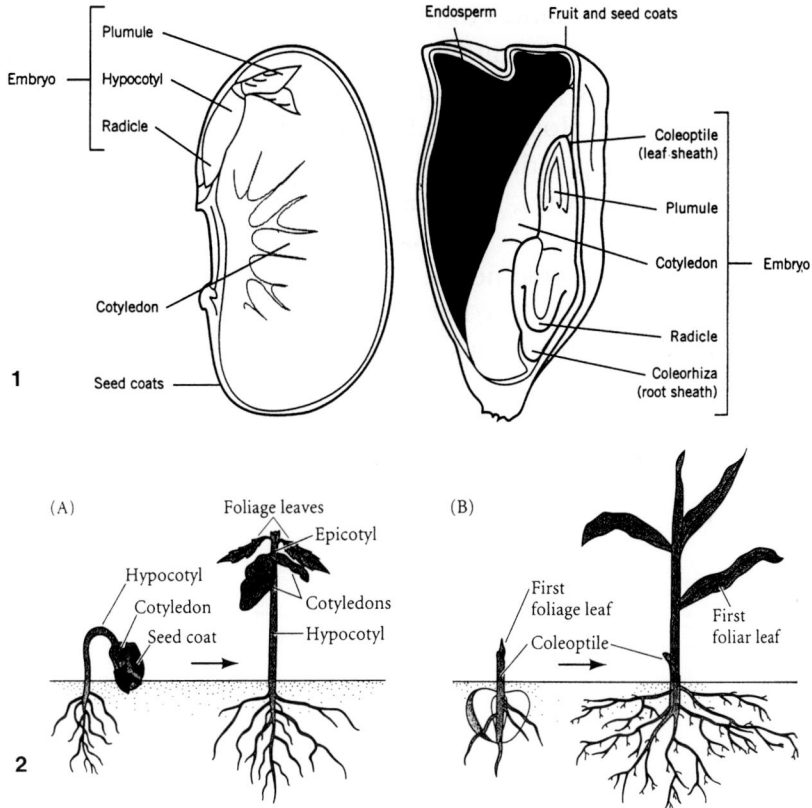

Fig. 2.5   The decision of what tissue will be growing in the direction of gravity, and will develop in the opposite direction, is made in advance in the embryo, before the plant emerges out of the seed.

1. Seeds of common bean, *Phaseolus vulgaris* (left) and maize, *Zea mays* (right) showing the embryo.

2. (A) and (B) developing plants of these species. In the embryo of the bean there are minute leaves (plumule) a hypocotyl that will become the stem, and a minor root (radicle). The embryo of maize contains a leaf sheath (coleoptile), plumule (that will become leaves) and a radicle encircled by a root sheath that will form the root system. Hence, the growth of the organs, in relation to gravity, is determined before the plant develops.

A simple experiment corroborates this evidence. Hormones move between tissues and organs within a plant. A unique characteristic of auxin transport is the polarity of movement. The polar transport is expressed as a preferential movement in one direction.

When the hormone moves to the apex of a stem or coleoptile this can be cut and the segment be inverted. What is surprising is that the inversion of the piece of stem does not affect the polar movement of the auxin. Hopkins and Hüner (2004) conclude that "this indicates that the polarity of the transport is not driven by external forces (e.g. gravity), but is an inherent property of the cells themselves".

The conclusion could not be more elucidating, the cells simply do not care for gravity. The hormones are sent up or down, left or right, according to the plant's own pattern of function.

22

Physicists Construct Antigravity Devices that Oblige Frogs and Plants to Float

The April 1997 issue of "Physics World", a reputable scientific journal, included the report of an antigravity device that was used to levitate a frog. The article was not a joke but was the product of the research of scientists led by Andrey Geim at Nijmegen University in Holland. The frog was floated using a powerful electromagnet. This result was a complete surprise even to leading physicists, but it represented a demonstration of a fundamental property of matter. Living tissues as well as water (that they contain in large amounts) are weakly magnetic, exhibiting what is called *diamagnetism*, which means that they have a slight tendency to become magnetized in a direction opposite to an intense magnetic field. The result is that an object with a small volume receives an upward force, counteracting gravity (Berry and Geim, 1997) (Fig. 2.6).

Simulated low gravity conditions could be easily obtained when objects were sent on space missions by NASA. Mark Meisel at the University of Florida, U.S.A., noted that the plant *Arabidopsis* (a mustard plant widely used in genetic experiments) showed signs of stress when submitted to periods without gravity. The experiments also included floating *Arabidopsis* seedlings with the help of powerful electromagnets, the reaction from the plants being similar to the zero-gravity environment (Schneider, 1999).

Fig. 2.6 A water droplet as well as a whole living organism, such as a frog, can be levitated floating freely when submitted in the laboratory to powerful electro-magnets.

1. Live frog and 2. water droplet float freely inside the 32-millimeter vertical bore of an elec-tromagnet at the University of Nijmegen in The Netherlands.

23

Atoms Hide Many Properties that May Disclose the Mechanisms Behind Living Processes — Liquid Helium Can Build a Fountain Ejecting Itself Out of a Flask

After hydrogen, helium is the most common chemical element in the universe. It is formed in stars by the fusion of hydrogen atoms at temperatures of 10 million degrees.

This chemical element remains in liquid form even at absolute zero temperatures and although it is a liquid it has a perfectly ordered structure.

Helium has the lowest boiling point of any gas and was the last to be liquefied. When this occurred it turned out to have many remarkable properties. These are due to the zero-point motion of the helium atoms since they are close to the absolute zero temperature of –273°C. The liquid helium can "creep" up the sides of a container, flow over the top and in other cases spout of a flask in a jet like a fountain (Hey and Walters, 2003) (Fig. 2.7).

Obviously we are dealing with extremely low temperatures far removed from those of living organisms, but this means that atoms hide many properties that may help in understanding their behavior in cells and chromosomes.

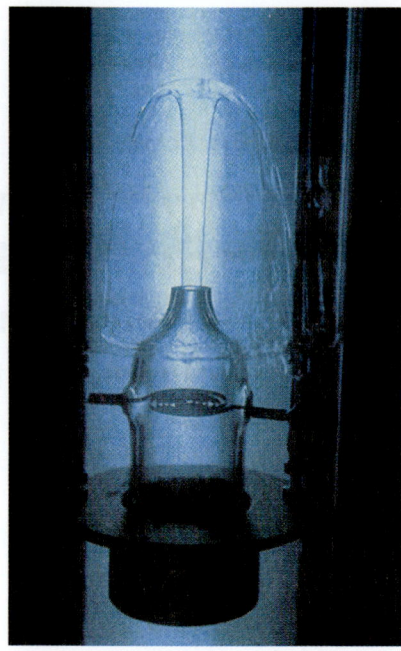

Fig. 2.7   A spectacular photograph of the "fountain effect."

The strange behaviour of liquid helium is evident. The helium ejects itself out of the flask. Its atoms condense into the lowest energy state interacting cooperatively, the result being the building of a super fluid.

24

Levitation in Metals — What was Impossible Became Possible

Physics is not such a developed science as we tend to believe. Obviously, when compared to biology it is most advanced, but when one starts to make inroads in several of its postulates they turn out to be incoherent or prone to modification.

Levitation in physics is one such area. Brandt (1989) has studied this phenomenon in detail. He recalls that: "In 1842 Earnshaw proved that stable levitation or suspension is impossible for a body placed in a repulsive or attractive static force field in which force and distance are related by an inverse square law. From Earnshaw's theorem one might conclude that levitation of charged or magnetic bodies is not possible, but this is not generally true." For many years Earnshaw's mathematical treatment of the phenomenon gave it a status of infallibility. At present, as Brandt points out, his theorem applies only to individual particles. In bodies with dipoles, stable levitation is possible in static magnetic fields. He adds then a series of situations in which levitation occurs.

Materials may be levitated by intense sound waves or by means of laser light. In addition superconductors may be suspended both above and below a magnet. Superconductors are metals cooled close to absolute zero temperatures which get a vanishing small electrical resistance. This novel property is due to changes in the arrangement of their electrons (Pitt, 1988). The levitation of superconductors is a most conspicuous phenomenon. It requires no

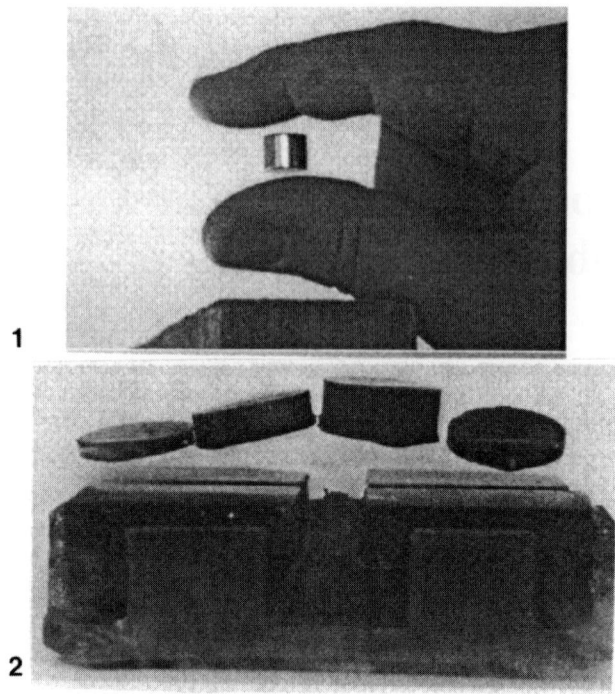

Fig. 2.8   Levitation of magnets and superconductors.

1. Levitation of a magnet located 2.5 m below an unseen superconducting coil of wire (solenoid) stabilized by the diamagnetism of human fingers. In diamagnetism, present in cell water and other substances, the electrons change their orbits and velocities so as to produce a magnetic field that opposes the applied field. The result is that the magnet becomes stably suspended between the fingertips.

2. Disks (12 mm in diameter) of the oxide superconductor $YBa_2\text{-}Cu_3O_7\text{-}8$ levitated above a permanent magnet with one central north pole and four south pole sections. A superconductor is a metal which has been cooled close to absolute zero and which has little electrical resistance.

energy input, it is static and stable (Fig. 2.8). Many tons may be lifted in this way, as has been demonstrated in Japan and Germany in the construction of high-speed trains.

What was impossible in 1842 became possible in our time, the mechanism being related to the particular arrangement of the electrons inside the material.

25

Animals, like Plants, Have Created Devices that Counteract Gravity

A beautiful and simple experiment can be carried out by any one having in his backyard a few hens and an active cock. The fertilized eggs can be collected and broken open at one day intervals following fertilization.

A unique spectacle becomes available as the homogenous yellow yolk of the egg suddenly builds a red blood patch. In its middle, by the second day, even before the shape of an embryo can be distinguished, there is already a tiny heart that beats with its own rhythm. It is life pulsating. You will never forget this encounter with a muscular force that will allow the heart to expand and contract, performing this task thousands of times, for years, without any intermediate stop or rest (Burton, 1987).

Before any other organ is visible there is a heart. The translocation of nutrients, necessary to produce the other organs, that make up the embryo, is ensured by this pumping system and a series of ramified vessels. This tiny heart becomes in the fully developed chicken a large and complex organ (Fig. 2.9).

At the very start of embryo formation, an organ is formed that disregards gravity. The blood is obliged to move in every direction throughout a system of vessels.

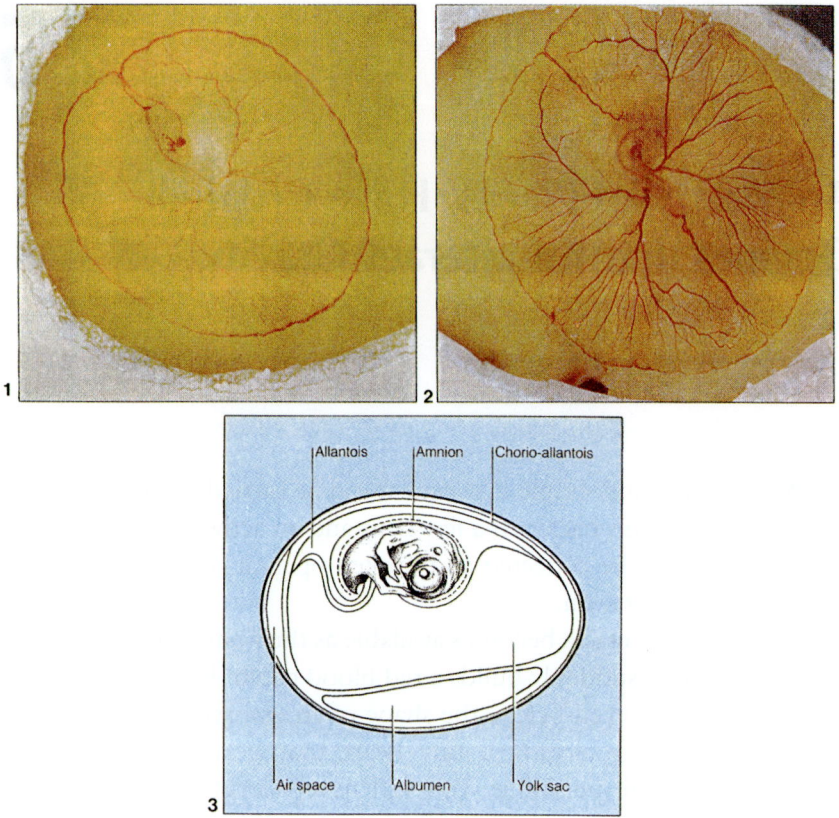

Fig. 2.9   In a hen's embryo the heart is visible before any other organ.

When a fertilized egg is broken open, after one or two days of development, a spreading disc of cells is seen sitting on the yolk.

1. Three days after fertilization the embryo has developed a network of blood vessels and a simple heart is pumping blood collecting much-needed nourishment from the yolk.

2. Nourished by the expanding blood system other internal organs take shape in the four-day-old embryo.

3. Diagram of a hen's egg at about ten days showing the well developed embryo.

26

In Giraffes the Distance between the Heart and the Head is Over Two Meters

The circulation created by the heart is actually counteracting gravity in even more extreme circumstances.

This phenomenon is evident in humans, who originated from apes walking on their four members, but who rose to an upright posture and the bipedal position. During this evolutionary transformation the heart was obliged to pump the blood to higher levels of the body including the head. A similar transformation occurs during the growth that humans undergo from childhood to adults. Again they rise, from a curved body posture at birth, to the adult upright position, by a series of curvatures that occur in their skeleton (Fig. 2.10).

But in animals, like the giraffes, the counteraction of gravity, takes its most extreme form. These mammals, which are only found on the African continent, are the tallest in the world. A mature male measures 5.3 m (17 ft) (Dorst and Dandelot, 1988; Burnie, 2004). The giraffe's great height means that, to drink water, it must splay its front legs and even bend at the knees. The head is then at the level of the soil. When the animal eats leaves at the top of trees, the head becomes located 5 meters above the ground. When upright, the heart of the giraffe has to pump blood upwards at enormous pressure to reach the brain at the top of a long neck.

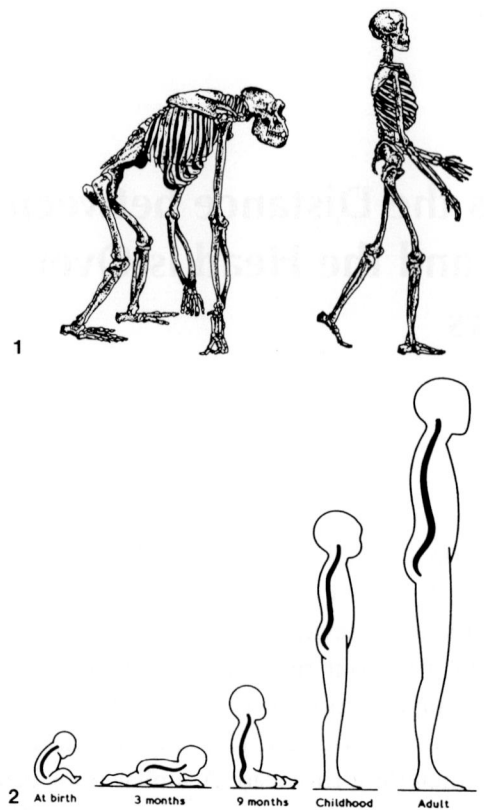

Fig. 2.10   The bipedal position of humans created novel blood movements.

1. Skeleton of *Gorilla* (left) and *Homo sapiens* (right).
2. Changes in the curvature of the human vertebral column during growth. The newborn possesses a single posterior curve like the quadrupeds such as the gorilla.

The distance between the heart and the head is over 2 m. Blood circulation in the giraffe has been studied extensively. Not only is the brain irrigated properly but, as the head is raised or lowered, the giraffe adjusts the blood pressure to compensate for the two extreme situations (Eckert and Randall, 1978) (Fig. 2.11).

For animals this is a great feat, but compared with the plants it means little. They pump their sap to 100 meters of height without the help of a heart or any other type of auxiliary organ. In simple

Fig. 2.11 A brain located 5 meters above ground can be easily reached by blood.
The giraffe is the tallest animal. It browses higher than any other mammal, mainly for aca-
cia leaves, thanks to its great elongated neck and long legs. The giraffe's male height is 4.7
to 5.3 m (15 to 17 ft) and the weight 800 to 1.930 kg (1.765–4.255 lb). The new born
weighs up to 70 kg (155 lb) and stands 2 m (6 1/2 ft) tall. The giraffe has a drinking problem. To drink water it must splay its front legs. When upright its heart has to pump blood
upwards at enormous pressure to reach the brain, but when the head lowers to drink, valves
regulate the blood flow preventing damage to the brain. *Inset* — A man compared to a
giraffe.

animals, like the invertebrates, such as the jellyfish and sea
anemones, there is no heart and the circulation of nutrients takes
place in a form apparently no different from that in plants (Russel-
Hunter, 1979).

These results teach us that, in living organisms, the same process
can be attained by quite different solutions.

27

No Chromosome Obeys Newton's Laws — In their Movements Chromosomes Bypass Gravity

O ver a hundred text books and monographs, dealing with chromosome research, have been published in the last 10 years. In them, the division and separation of chromosomes into daughter cells, is described in detail, as has been since these organelles were discovered in the period 1880–1900. However, the fact that the chromosome movements are highly peculiar and different from other types of movements occurring on the surface of the Earth seems to have escaped cytologists and geneticists (Lima-de-Faria, 2003). At every cell division, the chromosomes do not follow Newton's laws. As is well known, one chromosome set moves to one pole which may be in the direction of gravity, but the other set moves in the opposite direction. Again we are witnessing a biological system in which the action of gravity is invalidated. This event does not happen once, or in a particular tissue, but is of universal occurrence. Every time a plant or animal cell divides, its chromosomes may move following gravity or counteracting it, or in any other direction (Fig. 2.12).

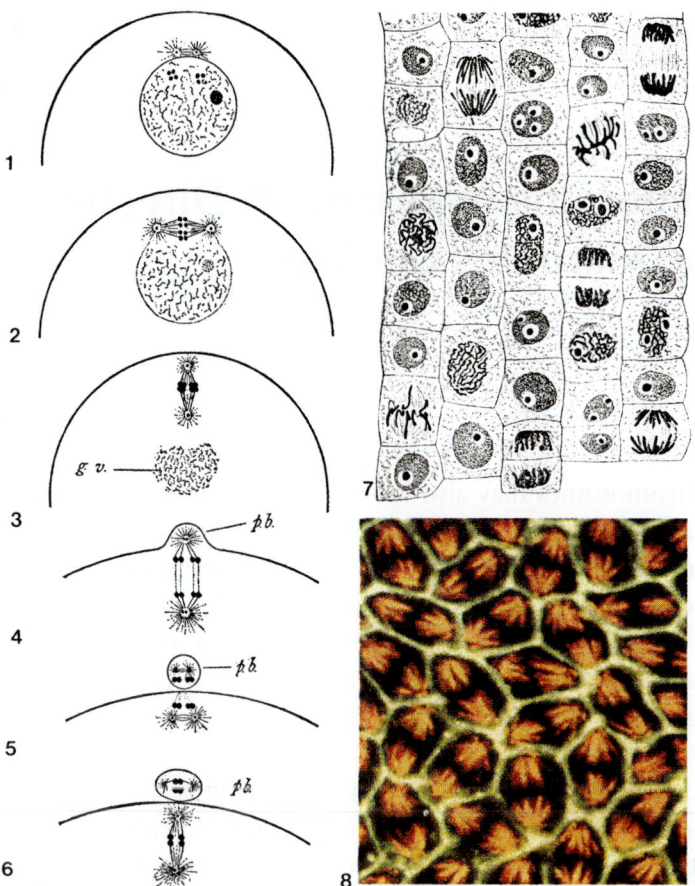

Fig. 2.12   Chromosomes divide in all possible directions I.

1–6. Alternating directions of chromosome movements in an egg with 4 chromosomes. (1) Two centrosomes, with their radiating asters, are located at the top of the nucleus and the chromosomes start dividing in a horizontal position (2). The centrosome-chromosome assembly changes to a vertical position leading the chromosomes to opposite poles (3). A protuberance is formed at the top of the egg, one set of chromosomes moves into it (4 and 5). A division occurs again in the horizontal direction within the "polar body" (p.b.). At the same time another division takes place vertically inside the egg (6). Hence, chromosomes, located side by side, are moving east-west and north-south.

7. Growing root-tip of an onion seen in longitudinal section.The chromosomes are separating from each other — a set moves in the direction of gravity the other set against the gravitational force. In some cases the movement is transversal.

8. Fluorescence photomicrograph of nuclei dividing in all possible directions in the embryo of the fly *Drosophila*. (Photograph courtesy of E. Theurkauf and W. Sullivan 2000).

28

Chromosomes Move in All Directions of the Mariner's Compass

Chromosomes may also move in particular ways. The direction of alternate nuclear divisions results in a zigzag pattern of chromosome arrangements as is the case during the segmentation of the egg in the insect *Wachtiella* (Fig. 2.13). In other eggs, in the first division, the chromosomes separate in opposite directions along a horizontal plane, but in the next division they separate along a vertical plane, to be followed by a third division along a horizontal plane (Fig. 2.12).

Chromosomes actually do not care for the position of the Earth, of the Moon or of the Sun. They may move in east-west or south-north direction. What they regularly do is that they, most of the time, separate in opposite directions as they move into daughter cells. This rule has some exceptions. There are situations in which small groups of chromosomes may move, within the same cell, in all the directions of the mariner's compass (Fig. 2.13).

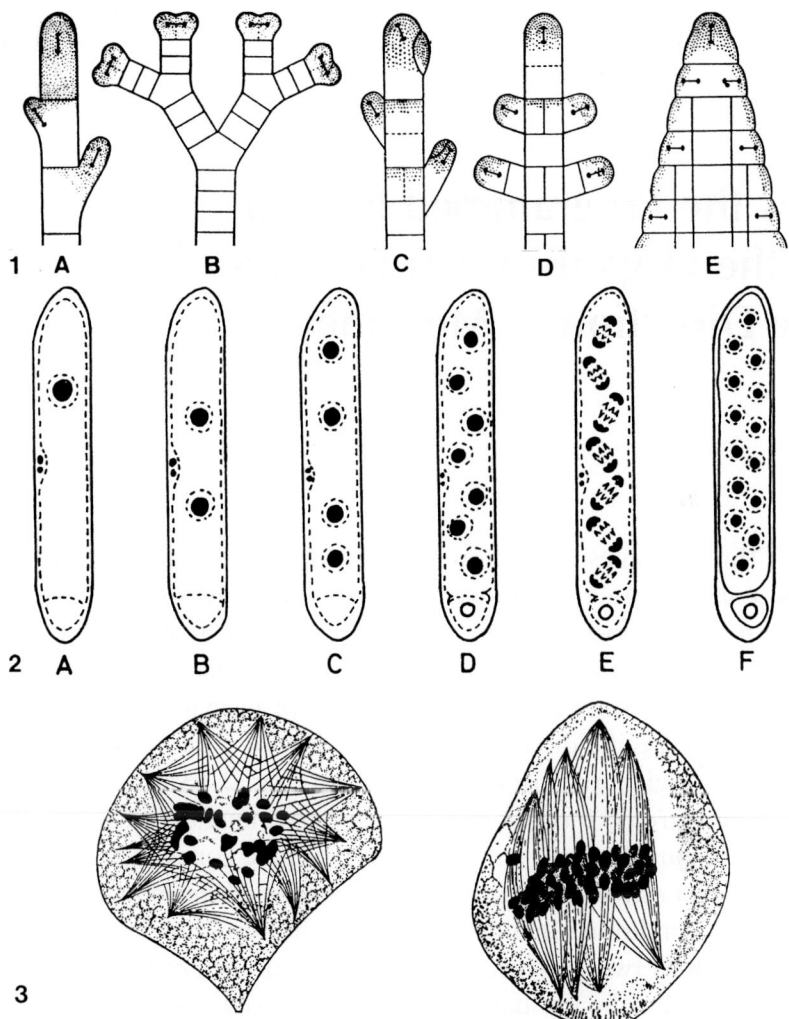

Fig. 2.13   Chromosomes divide in all possible directions II.

1. Types of branching of algae growing as flattened sheets. A. Filament growing from an apical cell showing apical-polar branching. B to E. Other types of branching with divisions taking place in different directions.

2. Segmentation of the egg in the insect *Wachtiella* (A to F) with criss-cross orientation of chromosome movements (E).

3. Division of two spore-mother cells in the fern *Equisetum*, showing multipolar spindle-formation.

29

It is the Programmed Pattern of the Organism that Decides the Direction of Movement

Chromosomes may show no regard for the earth attraction, since they actually move in any possible direction, but they are not free to choose their direction. This is rigidly determined by the architectural scaffolding consisting of the cells that are to form the organism, a feature evident in the growth of algae in which the cells branch in specific directions. The cell divisions are obliged to follow the original pattern of the organism (Fig. 2.13).

The extreme form of this pattern determination occurs in the cell divisions that lead to the building of pine cones in the conifer *Araucaria excelsa* and in the inflorescence of *Helianthus annuus* in fruiting condition. In this last species the resulting fruits can be divided into two groups. In one individual analyzed in detail 55 were found to be left-handed and 89 to be right-handed. In both cases the cell divisions are highly ordered, occurring in cell primordia, in such a way, that the resulting pattern approaches a spiral (Fig. 2.17). The rigid mathematical configuration discloses two features: (1) The sequence and the orientation of the cell divisions follow a well defined pattern that results in a cascade of events that is highly regular. (2) The same pattern and order, are repeated in every generation, obliging the same type of organism to be formed.

As we shall see soon the advanced determination of these patterns is canalized by specific RNAs and proteins.

30

The Devices Used by the Chromosome Which Result in Particular Movements

What are the devices that chromosomes possess that allow them such impressive movements? They are mainly two: Firstly, they are in the possession of a DNA sequence, called the centromere or kinetochore, that is mainly responsible for the movement of each chromosome. Secondly, long spindle fibers, as they are also called, occur in the cytoplasm. They consist mainly of proteins, such as tubulin. By their attachment to the centromere and subsequent movement, by shortening, the fibers guide the chromosomes to the spindle poles, rebuilding two nuclei with opposite positions, around which two new cells are formed. What an ordered process! Nothing is left to hazard in this most dynamic event otherwise it would not be possible to repeat it since the dawn of the cell type formed in the protozoa. This performance has led to an uncountable number of cell divisions.

The human body, which is not the largest among mammals, starts from a single cell (the fertilized egg) and ends up being a compilation of 10^{13} cells (Hood, 2002). The number of stars in the Universe is also an impressive figure but it may be smaller than the number of cell divisions that have been necessary to build all living organisms past and present.

The opportunity for errors has been tremendous but the chromosome's and the cell's organization has supplanted all difficulties to this day.

31

Chromosomes Move Inside the Nucleus, like Goldfish in an Aquarium, without the Use of Spindle Fibers

Usually, it is not mentioned that chromosomes have other types of movements than those involved in cell division.

The so called interphase, is the stage found between the end of a cell division and the initiation of the next. As the beginning of a new cell division approaches, the chromosomes become recognizable at the stage called prophase. At this time the chromosomes are imprisoned in the nucleus but are not passive and static structures. They are actually moving all the time. This phenomenon can be observed under the microscope in living cells, such as the staminal hairs of the small plant *Tradescantia,* an observation made already by Strasburger (1924) and Belar (1928).

When it became possible to make films through the lenses of the microscope the active movement of chromosomes became a spectacle to be watched by cytologists sitting in an armchair. Films were made in several plant species (Bajer, 1957). They showed the frenetic movements of the living chromosomes inside the nucleus. The chromosomes displaced themselves rapidly, forward and back, up and down, moving like goldfish in an aquarium. Remarkable is that, at this time, their centromeres are not actively participating in

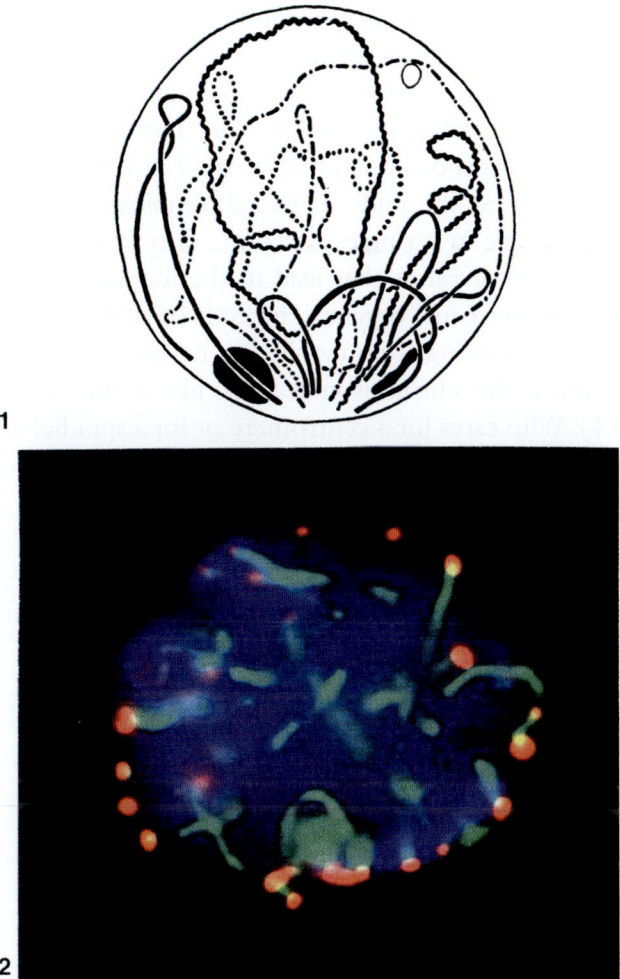

Fig. 2.14   The ends of chromosomes move inside the nucleus in the absence of spindle fibers and centrosomes.

1. Chromosomes in the insect *Chorthippus parallelus* with all their ends polarized building the so called "bouquet stage" (since it resembles a bunch of flowers).

2. Nucleus of a mouse cell at prophase of meiosis. The telomeres (stained red) of all chromosomes have moved to the nuclear envelope, locating themselves at the periphery of the nucleus. Image by courtesy of H. Scherthan, Munich, FRG.

the movement and there are no spindle fibers inside the nucleus. Hence, they move on their own.

One could think that they were jumping around due to Brownian motion, but this is not the case. There is also order during this stage. This is demonstrated by the fact that, in many plant and animal species, the ends of the chromosomes of the whole complement move in a single direction. This event occurs at the beginning of the divisions that lead to the formation of male and female gametes in sexual organs. The ends of all the chromosomes come together building what cytologists have called the "bouquet stage" because the chromosomes look like a bunch of flowers (Fig. 2.14). Who cares for a centromere or for a spindle! The chromosome is able to move orderly without them. Inside the nucleus the ends have already a guiding ability.

32

The Chromosome's Autonomy Takes Many Forms — The Chromosome Ends May Take Over the Function of Active Mobility on the Spindle

Cytologists discovered a new solution that the chromosome can display. Instead of the centromere leading the chromosome to the poles during the normal anaphase movement, the ends of these chromosomes functioned suddenly as neo-centromeres being highly active. Moreover, the two ends of the same chromosome could pull it in one direction, or each end could pull in opposite directions. This means that the DNA sequences of the end regions, which were substituting the function of centromeres, could also pull in different directions not caring the least for gravity. This phenomenon has been described by several authors in different plant species such as maize, rye, and wheat (Rhoades, 1952) (Fig. 2.15).

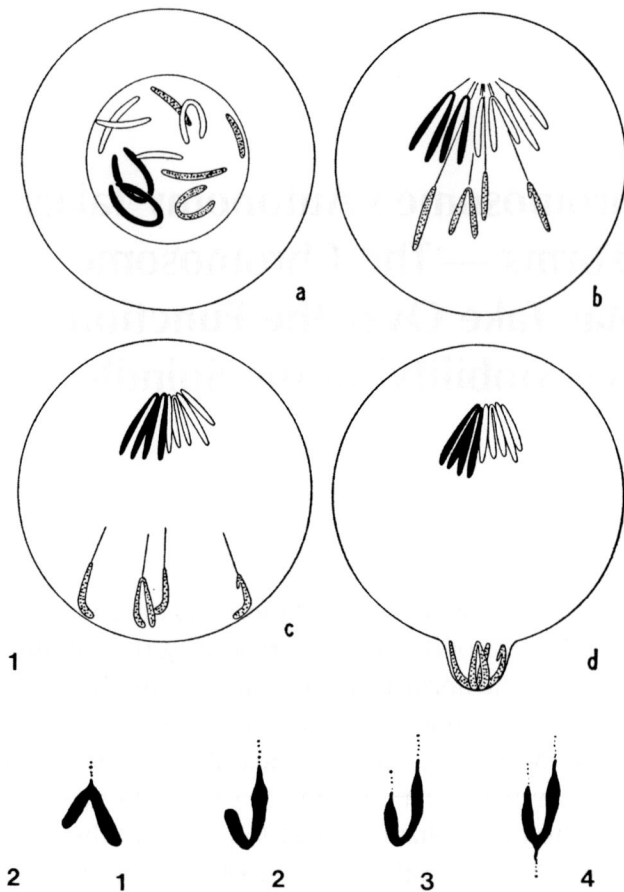

Fig. 2.15 How to get rid of the chromosomes from the father and how chromosome ends become active.

1. Divisions preceding the formation of sperm in the fly *Sciara* (a, b, c, d). Paternal chromosomes are stippled, maternal chromosomes are white and black. No pairing occurs, and the chromosomes do not assemble at the middle of the cell as is usually the case in other organisms. Instead, the maternal chromosomes move to the upper pole, whereas paternal chromosomes dash away into a bud from which they are cast off (d).

2. In maize chromosome ends take up the moving function from inactive centromeres. A normal chromosome with an active centromere leading the chromosome to the pole (1). Three cases in which the centromere is partly inactivated and instead the chromosome ends are moving actively (2, 3, 4).

33

Chromosomes Inherited from the Father May be Sent to a Different Compartment than those Inherited from the Mother

During the formation of sperm cells in the insect *Sciara coprophila* the spindle instead of being bipolar is cone shaped and unipolar. The whole chromosome complement becomes divided into two groups. One group moves to the single pole, whereas the other group, using the same spindle, is pushed in the opposite direction and finally eliminated. Genetic analysis has disclosed that the chromosomes that move as a group to the pole are the set derived from the mother whereas those that move in the opposite direction, and are later discarded, are those that were inherited from the father. Experimentally induced translocations between segments of these chromosomes have shown that their particular behavior was controlled by DNA sequences present in the chromosomes (Crouse, 1979). Thus, in *Sciara* the chromosomes are able to move backwards along the spindle without a spindle pole to go to, their strange movements being guided by the products of genes located in their chromosomes. Moreover, they have no difficulties to finally gather in a bud which is cast off (Fig. 2.15).

34

How to Move Equally Well without Guiding Asters and Centrosomes

In most pictures of cell division in animal cells, chromosomes are seen attached by their centromeres to spindle fibers that converge at the poles in a radial mass that has at the center a corpuscule. This radial mass, usually seen well in dividing eggs, is called the aster and the corpuscule from which they diverge, in a circular fashion, is the centrosome which contains two centrioles. Its complex structure has been analyzed with the electron microscope. The guiding roles of the aster and the centrosome in chromosome movements remain obscure, but they are supposed to contribute to the separation of daughter cells with the help of proteins (Fig. 1.2).

Of interest is that in flowering plants there is no centrosome and no aster, yet their chromosomes divide with equal efficiency and precision, as they do in animal chromosomes. Order prevails in plant cell division although what seem to be obligatory components of animal cell division can be easily dispensed with (Fig. 1.2).

35

DNA Replication Already Disregarded Gravity

The division of the chromosome into two daughter units is preceded by the replication of its DNA into two identical copies. When DNA replicates, it produces a copy of its own base sequence, but both the original and the copy have to separate in different directions, otherwise they cannot become independent.

If one followed the direction of gravity, the other had to move in the opposite direction. But they could even both move independently of gravity. What force was then more powerful than the all pervading attraction to the earth. The main source of direction is in the atomic forces present in the nucleic acid molecules. These ignore gravity due to their stronger electromagnetic power.

The copying mechanism of DNA consists of many steps. As Brown (2007) points out the initiation of replication is "not a random process". It always begins at specific positions in the DNA molecule which are as many as 20,000 in the human chromosome complement. A collection of proteins orchestrates the event. These are the topoisomerases that introduce breaks in the DNA allowing the unwinding of the helices. Other proteins help in their separation such as the helicases.

The DNA molecule consists of a double helix that is held together by hydrogen bonds between specific pairs of bases. Hydrogen bonds convey electrostatic attractions that have the special property of being at the same time strong and weak. They are strong enough to keep the DNA helices together, but when their holding capacity diminishes, their weakness results in the separation of the two DNA strands. Their alternate behaviour is determined by the chemical composition of the cell environment (King and Stansfield, 1997).

Hence, it is not gravity, but the electrostatic attractions between atoms, combined with the intervention of specific proteins, that decide the separation of the daughter units of DNA.

36

Mineral Crystallization Disposes Also of Gravity and Imposes Massive Copying

If you enter your kitchen and dissolve in tap water a large quantity of common salt (sodium chloride) and place it on a flat dish, the water soon evaporates and solid crystals of salt are formed. These do not grow as a flat carpet, on the base of the dish, but move upwards in different directions producing impressive pyramids (Fig. 2.16). As the atoms of sodium chloride, are liberated from the water, they assemble together with such a force that they counteract gravity. This force is the one that unites atoms into molecules and that physicists call the "electromagnetic force" because it glues atoms together with high efficiency. The same occurs in the crystals of the mineral aragonite (calcium carbonate) and those of gypsum, which grow in all possible directions without regard to gravity (Fig. 2.16).

The strength of this interaction between the atoms is even better demonstrated by the experiments made with such a simple chemical as brown sugar. When small crystals of this sugar are introduced into a sugar solution they act as "seeds" initiating a general crystal growth. What is remarkable, and Davey (2004) calls it "miraculous", is that the initial small crystals have a mass that was only 1% of that of the final crystals obtained. What happens is that

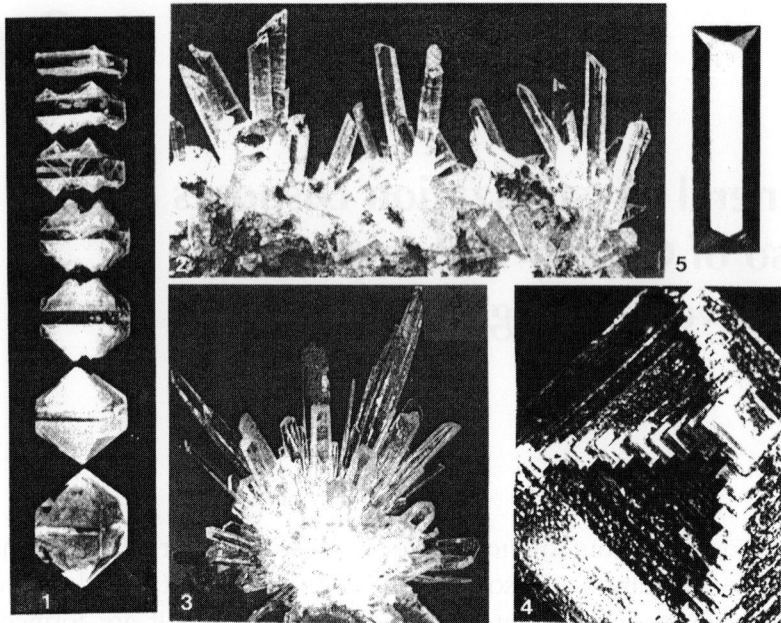

Fig. 2.16 The growth of crystals does not obey gravity.

1. Successive stages of the growth of a crystal of ammonium dihydrogen phosphate. It starts with a flat sheet and grows by adding atoms on each side, building two pyramids, one facing gravity the other opposing it.

2. Crystals of gypsum and 3. crystals of aragonite growing in all directions. 4. Crystals of common salt (NaCl) building a pyramid. 5. Photomicrograph of a crystal of the protein hexokinase, a key enzyme in the utilization of glucose (Courtesy of T. Steitz and M. Yeager).

the initial pattern is rapidly transmitted and imposed on a number of atoms many times larger, leading to a massive copying. It is thus not surprising that the atoms in DNA and in the chromosome behave accordingly.

37

The Egg Contents Rotate and the Cells Move within the Embryo Independently of Gravity

The unfertilized egg of frogs has a polarity along its main axis. This is clearly visible with the naked eye. This polarity is responsible for the mapping of tissues in the future embryo and it occurs before fertilization.

As the sperm fertilizes the egg there is a rotation of the molecular components of the cytoplasm. This rotation is upwards, directed towards the site of sperm entry, the contents moving 30 degrees in relation to the main axis. This upward movement is against gravity, and if the rotation of the molecular mass is blocked by artificial means, the embryo stops developing, looses its pattern becoming a formless mass, and finally dies (Vincent *et al.*, 1986).

At later embryonic stages a similar event can be observed. This time it is whole groups of cells that move upwards. The cells can actually be seen crawling up the inner surface of the embryo towards the animal pole. These cell movements against gravity are an obligatory step in the normal development of the organism, as demonstrated by several experiments (Holtfreter, 1944; Wilson and Keller, 1991; Winklbauer and Schürfeld, 1999).

Hence, cell contents, and groups of cells, move independently of gravity reaching specific sites. If their movement is arrested or blocked, the pattern of the embryo is eliminated. The counteraction of gravity is necessary for the emergence and maintenance of order.

The general pattern map of the embryo is known to be dependent on specific proteins which are produced by maternal messenger RNAs. This was experimentally demonstrated by Zhang *et al.* (1998) by depleting early embryos of these proteins. The result was a loss of the normal pattern map. Hence, the original source of the guidance of these movements, goes back to the RNAs produced by the chromosomes of the mother.

38

Snails Have a Shell in the Form of a Spiral Which is Either Left-Handed or Right-Handed — The Choice is Determined by Genes Which Orient the Axis of Cell Division Independently of Gravity

The body of snails consists of a shell in which spiral coils open on the right in most species. In rare cases the shell coils open to the left.

The determination of the two alternative patterns occurs already at the egg's second cell cleavage. The mitotic spindle orientation at this initial stage is responsible for the type of coiling of the future organism. All cell divisions that occur after this event in left-coiling animals are mirror images of those taking place in the right-coiling embryos. The change in direction of the spindle orientation occurs so early, because it is not directed by the snail's own chromosomes, but by a single pair of genes present in the snail's mother whose products are carried into the egg's cytoplasm (Fig. 2.17).

Experimental demonstration was obtained by Freeman and Lundelius (1982) who injected cytoplasm from right-handed coiling snails into the eggs of left-handed mutants transforming them into right-handed individuals. The change in orientation is due to

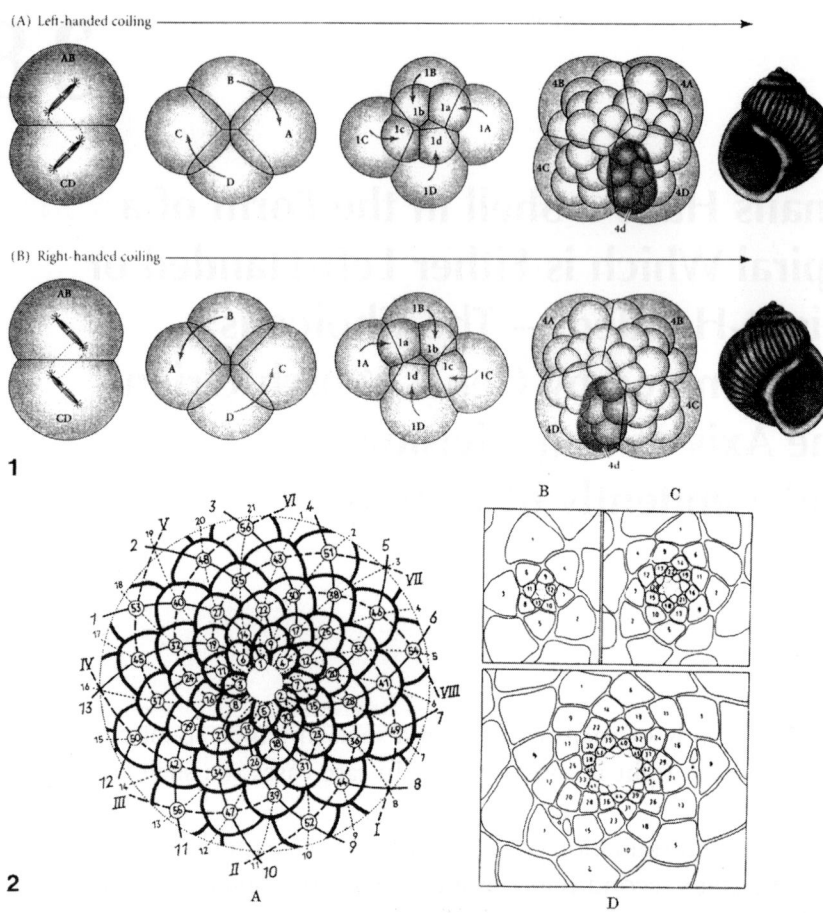

Fig. 2.17 Left-coiling and right-coiling snails, as well as left-handed and right-handed spirals in plants, are the product of autonomous chromosome movements.

1. A view from above of the animal pole of eggs of left-coiling and right-coiling snails. The origin of sinistral and dextral coiling can be traced to the orientation of the mitotic spindle at the second cleavage of the egg. The initial chromosome movements occur in opposite directions independently of gravity. The left-coiling (A) and right-coiling (B) snails develop as mirror images of each other.

2 A. Diagrammatic view of a pine cone from below. The cone scales are arranged in clockwise and anticlockwise spirals.

2 B, C and D. Transverse sections through three apices of the conifer *Araucaria excelsa*. B, right-handed spiral; C and D, left-handed spiral.

proteins in the cytoplasm interacting with the actin of the cell's cytoskeleton.

The spiral pattern is so simple, because it can be defined mathematically, and at the same time is so complex since it stands for the shape of the whole body of the organism. Inherited chromosome products, are actually determining a whole organism's pattern that has no direct relationship with gravity.

39

It Comes as a Revelation that the Chromosome Does Not Need Strong Magnets and Extreme Low Temperatures to Evade Gravity

The chromosome uses spindle fibers to move to the cell poles during its division. One could expect that it used the spindle apparatus only when it needed help to counteract gravity. But no, the chromosome uses the spindle, irrespective of the direction that it chooses. More significant, is that it moves equally well when no spindle is available as is the case when it turns around inside the nucleus in many directions. Besides, in animals asters and centrosomes are involved in chromosome movements but in plants they are not needed. Hence, we are dealing with a full autonomy of behavior. What comes forward as a revelation is that the frog needs powerful magnets, and helium demands extremely low temperatures, to evade gravity. But the chromosome does it without any strong magnets and at room temperature. Moreover, it performs this trick all the time and irrespective of cell type and organism complexity. This ability has remained hidden to us because we have not seriously considered its own power of innovation and independent behavior. Suddenly, what appears, as folly and impertinence, may shine allowing us to understand its unique behavior.

40

Goddesses Do Not Obey
Earthly Laws

It may be added that chromosomes do not seem to be much different from the products of human imagination, such as the gods and goddesses of Greek and Roman mythology. The gallery that they formed was most diversified. Athena, the goddess of wisdom, was born from the head of Zeus when it was split open with one blow, but Aphrodite, the goddess of love (who the Romans called Venus), rose from the sea and came ashore at the island of Cyprus. Love, for the Greeks, was synonymous with life and it is thus natural that her birth was unconventional — the ocean being seen as the source of all life. The Italian artist Sandro Botticelli (1444–1510) made her birth still less conventional. He started by painting the goddess naked. For thousand years a woman had not been painted totally naked because the Christian Church had banned the female nude considering it taboo. Moreover, Botticelli decided that she should not obey gravity. All her weight falls on a single foot located at the edge of a large shell that is supposed to float on a calm sea. Obviously, goddesses do not need to obey earthly laws (Fig. 2.18).

Fig. 2.18   A goddess does not obey gravity.

The "Birth of Venus" was painted by the Italian artist Sandro Botticelli (1444–1510) about 1485. What contributes to make it a masterpiece is that every detail in this painting defies convention. Like most Botticelli women she has a distressed and distracted looking. Moreover, Venus is not born from another goddess's womb but arises from the ocean, the source of all life, being carried to the shore among spring flowers blown by gentle winds. All her weight falls on a single foot located at the edge of a large shell that is supposed to float on a calm sea. The painting conveys a sense of weightlessness and dignified beauty reserved for goddesses.

References

Part II

Webster N. (1976) *Webster's New Twentieth Century Dictionary*. Collins World, USA.

David I, *et al.* (2002) *The Cambridge Dictionary of Scientists*. Cambridge University Press, Cambridge, UK.

Dicke RH. (1981) Gravitation. In: Lerner RG & Trigg GL (eds), *Encyclopedia of Physics*, pp. 364–367. Addison-Wesley Publ Co, Reading, MA, USA.

Hey T, Walters P. (2003) *The New Quantum Universe*. Cambridge University Press, Cambridge, UK.

von Baeyer HC. (1992) *Taming the Atom*. Random House, NY, USA.

Pitt VH. (1988) *The Penguin Dictionary of Physics*. Penguin Books, London, UK.

Glashow S. (1997) Blessed is the weak. *New Scientist*. October 1997: 28–31.

Pagels HR. (1982) *The Cosmic Code. Quantum Physics as the Language of Nature*. Michael Joseph, London.

Clark RW. (1973) *Einstein. The Life and Times*. Hodder and Stoughton, London.

Hogan J. (2007) Physicists plan search for the known unknowns. *Nature* **445**: 468–469.

Dawson TE. (1988) Fog in the California Redwood Forest: Ecosystem inputs and use by plants. *Oecologia* **117**: 476–485.

Woodward I. (2004) Tall storeys. *Nature* **428**: 807–808.

Hopkins WG, Hüner NPA. (2004) *Introduction to Plant Physiology.* Wiley, USA.

Richardson M. (1975) *Translocation in Plants.* Edward Arnold, London, UK.

Bidwell RGS. (1979) *Plant Physiology.* MacMillan Publ Co, NY, USA.

Berry MV, Geim AK. (1997) Of flying frogs and levitrons. *Eur J Physics* **18**: 307–313.

Schneider D. (1999) Some levity in physics. *American Scientist* **87**: 122–123.

Brandt EH. (1989) Levitation in physics. *Science, New Series* **223**: 349–355.

Burton R. (1987) *Egg, Nature's Miracle of Packaging.* William Collins, London, UK.

Dorst J, Dandelot P. (1988) *The Larger Mammals of Africa.* Collins, London, UK.

Burnie D. (2004) *Animal.* Dorling Kindersley, London, UK.

Eckert R, Randall D. (1978) *Animal Physiology.* Freeman, San Francisco, USA.

Russel-Hunter WD. (1979) *A Life of Invertebrates.* Macmillan Publishing Co, NY, USA.

Lima-de-Faria A. (2003) *One Hundred Years of Chromosome Research and What Remains to be Learned.* Kluwer Academic Publishers 2003, Dordrecht, London and 2004 Springer, NY, USA.

Hood L. (2002) After the genome, where should we go. In: Yudell M & DeSalle R (eds), *The Genomic Revolution*, pp. 64–73. Joseph Henry Press, Washington DC, USA.

Strasburger E. (1924) *Handbook of Practical Botany.* Allen and Unwin, London, UK.

Belar K. (1928) *Die Cytologischen Grundlagen der Vererbung.* Berlin, Germany.

Bajer A. (1957) Ciné-micrographic studies on mitosis in endosperm. III. The origin of the mitotic spindle. *Exp Cell Res* **13**: 493–502.

Rhoades MM. (1952) Preferential segregation in maize. In: JW Gowen (ed), *Heterosis*, pp. 66–80. Iowa State College Press, Ames, Iowa, USA.

Crouse HV. (1979) X heterochromatin subdivision and cytogenetic analysis in *Sciara coprophila*. II. The controlling element. *Chromosoma* **74**: 219–239.

Brown TA. (2007) *Genomes 3*. Garland Science, NY, USA.

King RC, Stansfield WD. (1997) *A Dictionary of Genetics*. Oxford University Press, Oxford, UK.

Davey RJ. (2004) How come you look so good? *Nature* **428**: 374–375.

Vincent JP, *et al.* (1986) Kinematics of gray crescent formation in *Xenopus* eggs. *Dev Biol* **113**: 484–500.

Holtfreter J. (1944) A study of the mechanics of gastrulation. Part II. *J Exp Zool* **95**: 171–212.

Wilson P, Keller R. (1991) Cell rearrangement during gastrulation of *Xenopus*. *Development* **112**: 289–300.

Winklbauer R, Schürfeld M. (1999) Vegetal rotation, a new gastrulation movement involved in the internalization of the mesoderm and endoderm in *Xenopus*. *Development* **126**: 3703–3713.

Zhang J, *et al.* (1998) The role of maternal Veg T in establishing the primary germ layers in *Xenopus* embryos. *Cell* **94**: 515–524.

Freeman G, Lundelius JW. (1982) The developmental genetics of dextrality and sinistrality in the gastropod *Limnea peregra*. *Wilhelm Roux Arch Dev Biol* **191**: 69–83.

Sources of Illustrations

Part II

2.1   Picasso E. (1996) Un grande progetto Europeo: L'anello di colli-sione electrone-positrone. *Atti Accademia Nazionale dei Lincei, Rome, Supplemento, serie 9, 7*: 3–22. (Fig. 1, page 5).

2.2   Schopper H. (1982) The search for the basic laws of the cosmos. CERN-14 1982. *Sudetendeutsche Beiträge zur Naturwissenschaft und Technik*, Bavarian Academy of Sciences, Munich, Germany, 1980. (Fig. Penetration into deeper layers of matter, page 2).

2.3   Richardson M. (1975) *Translocation in Plants*. Edward Arnold, London. (Fig. 3–3(a) and Fig. 3–3(b), page 20).

2.4   Taylor TN, Taylor EL. (1993) *The Biology and Evolution of Fossil Plants*. Englewood Cliffs, Prentice Hall, USA. In: Willis KJ, McElwain JC. (2002) *The Evolution of Plants*. Oxford University Press, Oxford, UK. (Fig. 4.15, page 103).

2. 5   (1)   Hopkins WG, Hüner NPA. (2004) *Introduction to Plant Physiology*. John Wiley and Sons, Inc., USA. (Fig. 1.30, page 23).

       (2)   (A and B) Gilbert SF. (2000) *Developmental Biology* (Chapter by Singer SR) Sinauer Associates, Sunderland, MA, USA. (Fig. 20.19, page 634).

2.6   (1) and (2) Schneider D. (1999) Some levity in physics. *American Scientist* **87**: 122–123. (Fig. Water droplet and live frog, page 123).

2.7   Hey T, Walters P (2003) *The New Quantum Universe*. Cambridge University Press, UK. (Fig. 7.13, page 143).

2.8 (1) Simon MD, Geim AK. (2000). Diamagnetic levitation: Flying frogs and floating magnets (invited). *J Applied Physics* **87**: 6200–6204. (Fig. 2, page 6201).

(2) Brandt EH. (1989) Levitation in physics. *Science New Series* **243**: 349–355. (Fig. 1, page 349).

2.9 (1), (2) and (3) Burton R. (1987) *Egg, Nature's Miracle of Packaging.* William Collins, London, UK. (Fig. A hen's embryo and Fig. Diagram of a hen's egg, page 107).

2.10 (1) Savage RJG, Long MR. (1986) *Mammal Evolution, An Illustrated Guide.* British Museum, London, UK. (Fig. on page 237).

(2) Napier JR, Napier PH. (1985) *The Natural History of the Primates.* British Museum, London, UK. (Fig. 3.31, page 51). Redrawn from Harrison RJ, Montagna W (1969) *Man.* Appleton-Century-Crofts, New York, USA.

2.11 Burnie D. (2004) *Animal.* DK, Dorling Kinder Sley, London, UK. (Fig. Giraffe and young, page 242).
Inset — Macdonald D. (1984) *Encyclopedia of Mammals: 2.* George Allen and Unwin, London, UK. (Fig. Giraffe and human, page 534).

2.12 (1 to 6) Wilson EB. (1925) *The Cell in Development and Heredity.* The Macmillan Co, New York, USA. (Fig. 237, page 499).

(7) Wilson EB. (1925). *The Cell in Development and Heredity.* The Macmillan Co, New York, USA. (Fig. 2, page 5).

(8) Gilbert SF. (2000) *Developmental Biology* (sixth edition). Sinauer Associates Inc Publishers, Sunderland, MA, USA. (Fig. 9.3, page 265). Photograph courtesy of E Theurkauf and W Sullivan (2000). Sullivan W, *et al.* (1993). *Development* **118**: 1245–1254.

2.13 (1) Denffer von D, *et al.* (1971) *Strasburger's Textbook of Botany.* Longman, New York, USA. (Fig. 71, page 75).

(2) Geyer-Duszynska I. (1959) Experimental research on chromosome elimination in *Cecidomyidae (Diptera). J Exp Zool* **141**: 391. From Lewis KR, John B. (1963) *Chromosome Marker.* J and A Churchill, London, UK. (Fig. 78, page 203).

(3) Osterout WJV. (1922) *Injury, Recovery and Death.* Lippincott, Philadelphia, USA. From Wilson EB (1925) *The Cell in Development and Heredity.* Macmillan, NY, USA. (Fig. 65, page 152).

2.14 (1) Darlington CD. (1937) *Recent Advances in Cytology.* J and A Churchill Ltd, London, UK. (Fig. 22, page 91).

(2) Scherthan H. (2001) A bouquet makes ends meet. *Nat Rev Mol Cell Biol* **2**: 621–627 (Fig. 2F, page 625).

2.15 (1) Schrader F. (1953) *Mitosis.* Columbia University Press, NY, USA. (Fig. 19, page 107). From Metz CW (1933) *Biol Bull Woods Hole,* USA **54**: 333–347.

(2) Rhoades MM. (1952) Preferential segregation in maize. In: JW Gowen (ed), *Heterosis,* pp. 66–80. Iowa State College Press, Ames, Iowa.

2.16 (1) Egli PH. (1949) Crystal research. *Sci Mon* **68**: 270–278.

(2) Cabrera A. (1937) *Historia Natural, Geologia,* 4. Instituto Gallach, Barcelona, Spain (Fig. on page 173).

(3) Medenbach O, Sussieck-Fornefeld C. (1983) *Minerais* (Translation of Mineralien, Mosaik Verlag, Munich, 1982). Ed Publica, Lisbon, Portugal. (Fig. Aragonite, page 141).

(4) Ehrhardt A. (1939) *Kristalle.* V Heinrich Ellerman, Hamburg, Germany. (Fig. Steinsalz, page 49).

(5) Stryer L. (1981) *Biochemistry.* Freeman and Co, NY, USA. (Fig. 2.1, page 11, Courtesy of Dr Thomas Steitz and Dr Mark Yeager).

2.17 (1) Gilbert SF. (2000) *Developmental Biology.* Sinauer Associates Inc, Sunderland MA, USA. (Fig. 8.28, page 242). After Morgan TH. (1927). *Experimental Embryology.* Columbia University Press, NY, USA.

(2) Denffer D von, *et al.* (1971). *Strasburger's Textbook of Botany.* Longman, London, UK. (Fig. 137, page 127).

2.18 Studenbrock C, Töpper B. (1999). *1000 Mästerverk i den Europeiska Bildkonsten* (Swedish translation). Könemann, Gütersloh, Germany. (Fig. on page 111).

Who Cares for Randomness

41

Randomness Was Originally Foreign to Science

An event that is random "happens without a definite plan, pattern or purpose" (Collins, 1987) and is "without direction, rule or method" (Webster, 1976). Randomness is the process in which random is supposed to occur, being a synonym of chance.

Babylonian, Egyptian, Greek and Roman science were based on a search for order. For it was only the discovery of order that allowed prediction of phenomena.

Babylonian astronomy started with the discovery of the motion of planets, which allowed to foresee their movements and eclipses. Egyptian geometry led to the exact construction of the great pyramids. They were so rigorously planned that they have stood for over 3,000 years. Greek mathematicians established the relationship between the length of a vibrating string and its pitch, uniting physics with music. Roman scientists, such as Pliny the Elder (A.D. 23–79) wrote "Naturalis Historia" consisting of 37 books, which was an encyclopaedia of all contemporary knowledge on animals, plants and minerals. Such a classification was a first step to introduce order into natural science by codifying nature.

Moreover, all ancient science was a search for mathematical coherence, as expressed in the establishment by Pythagoras (572–497 B.C.) of the relationships between the sides of a triangle and in the principles of Archimedes (287–212 B.C.) that allowed the measurement of volumes and weights.

The amount of trade during antiquity was small compared to what it became later. Cereals were the main source of food, and wine was part of commerce. Gold and silver were the most important precious metals but their quantities were also relatively small. Economic speculation occurred, but was of minor dimensions and gambling was practically absent.

The use of dice, in the form of small bones, was used by women. But the main objective of their use was divination of future events. This activity was essentially religious, not being part of gambling or of science. However, the concepts of random and chaos were not totally foreign to antiquity. But their existence was considered the opposite of order, and as a consequence they had hardly any place in the science of the time, which so successfully discovered the principles that governed the shapes of geometric figures, the movements of celestial bodies and the basic organization of living organisms.

As Honour and Fleming (2002) have pointed out, the words inscribed on the entrance of Plato's Academy were: "Let no one enter here who is ignorant of geometry". Geometry is the ultimate concretization of ordered relationships.

Randomness is an Economic Concept Introduced into 17th Century Science

The statement by Bynum *et al.* (1983) in their Dictionary of the History of Science is elucidating: "There exists surprisingly little evidence of any interest in the concept of probability before the mid-17th century". And they add "Although people had always gambled often using sophisticated randomizing devices such as dice, cards and lots, the study of the mathematics of such things began seriously only with J. Cardano (1501–1576) and Galileo (1564–1642) who wrote briefly about gaming problems".

The emergence of the interest, and the dates given, coincide with the beginning of the explosive trade that followed the discovery of the sea routes from Portugal and Spain to India (1487–1498) and the American continent (1492–1500). Moreover, the exploration voyage of Ferdinand de Magellan (1519) across the Pacific Ocean, demonstrated that the earth was round. This event had an economic and scientific impact. The stars of the southern sky were chartered. One could, for the first time lay one's hand over two globes: one for the whole earth and another for the whole sky. It was a golden period for geographers and astronomers as well as merchants (Fig. 3.1).

This was the time when large boats loaded with spices and other precious materials returned to European ports. Their journeys had

Fig. 3.1   Putting one's hand over the spherical globe.

The exploration voyage of Ferdinand de Magellan (1519) demonstrated that the earth was round and that India could be reached by navigating along the coast of Africa but also across the Pacific. Moreover, the stars of the southern sky were described in detail for the first time. It became a golden period for geographers and astronomers. One of them, is seen in this painting of 1668 by the Dutch artist Jan Vermeer (1632–1675) putting his hand on the new globe.

to be financed and their cargo brought rapidly great wealth. At the same time many of these ships never came back due to storms and warfare. Suddenly, large sums of money were invested whose fate was highly uncertain. Much could be gained but equally much could be lost. Hazardous investment and gambling became the order of the day (Figs. 3.2 and 3.3).

Fig. 3.2 The discovery of the sea routes to india and the American continent led to an explosion of trade in Europe.

Lisbon harbour in the early 16th century from an etching in Bry's book "America". The rising sun that dominates the picture with its many rays, is an expression of the economic and scientific optimism that reigned at the time. From 1480 to 1580 Lisbon was the mercantile center of Europe. Ships not only arrived loaded with spices and other merchandise, but brought new plants and animals never seen before. One of them was a living rhinoceros offered to the King of Portugal by an Indian sovereign that was later painted by the German artist Albrecht Dürer (1471–1528).

The trade expansion continued, and by the 17th century, other European nations, such as Holland, England and France, joined with large merchant fleets.

It is then that B. Pascal (1623–1662), P. Fermat (1601–1665) and C. Huygens (1629–1695) wrote on mathematical probability between 1654 and 1657 (Katz, 1993). Pascal went one step further and constructed the first calculating machine (1652). In this period several women became leading scientists and it was they who taught mathematics to educate men (Fig. 3.4).

During the two following centuries physics made extensive use of probability, due to difficulties in predicting the behavior of

Fig. 3.3   Gambling became a quick way to get money for both men and women.
The French artist Georges de La Tour (1593–1652) called his painting appropriately: "The Cheat (with the Ace of Diamonds)" 1635. The man to the left holds behind him two cards, one of them the Ace of Diamonds, to cheat the woman in the center whose eyes express her distrust. Money can be seen lying on the table (to the right). Gambling was so widespread at this time that several books were published dealing with its moral implications.

atoms and molecules in gases and liquids. By the early 20th century this methodology entered biology mainly in connection with the study of animal and human populations. Statistics soon became one of the favorite tools of genetics as the laws of inheritance were formulated in statistical terms and the evolution of populations was assumed to be due to random mutation.

Fig. 3.4   Women teach mathematics to men.

During the Renaissance women acquired a leading position in society, many of them becoming teachers and prominent scientists. In this tapestry from the 1400s a lady points with one hand to a book and with the other to 15 pieces that enable rapid calculation. A man is her student and, over her head, is written *Aritmetique* the French word for arithmetic.

43

Randomness is Synonymous with Ignorance — "The Folly of Probability"

Following in the tradition of Pascal, the French physicist and mathematician P.S. Laplace (1749–1827), was the principal developer of probability theory. But he warned against the tendency to lean on randomness. For Laplace "The term randomness expresses only our ignorance of the causes of the phenomena that we see emerging in front of us and to succeed each other without apparent order" (Rebière 1900).

Such a cautious attitude had little effect on biologists for a long time. One such voice was that of the British cytologist C. Darlington (1953), who although he established, in a brilliant way, the predictable behavior of chromosomes, was led to point out the sources of uncertainty which dominated cell research. For him, like for most of his colleagues at that time, there were two main chromosome events which were random processes: mutation and chromosome recombination. They arose through a "multitude of accidents" (Darlington, 1953). Many probability analyses were published to explain the role of mutations in evolution or the origin of the cell. Computer calculations were also used to describe the origin of the chains of proteins, such as hemoglobins. All studies had to assume billions of cell generations and astronomic numbers of mutations, which exposed their inability to deal with these phenomena (Eden, 1967).

Soon other voices pointed in another direction. The American chemists L. Pauling and E. Zuckerkandl (1972) clearly rejected the introduction of chance in evolution. For them "Chance is never an intrinsic quality of things, it is an expression of the situation of the observer with respect to the observed". Few listened to their argumentation.

The misuse of randomness in biology and genetics is what Mora (1965) called "The folly of probability". This folly is best understood when one considers the emergence of proteins in connection with the origin of the cell.

Ribonuclease is one of the smallest known enzymes, containing only 124 amino acid residues. If the first organism contained an enzyme still smaller than this, e.g. constituted of 100 amino acid residues, and if this enzyme had been formed by chance this would mean one out of 1.3×10^{30} possibilities. Haldane (1965) thought that this was an impossible event. He calculated that if an organism were to be tried out every minute for 10^8 years it would necessitate 10^{17} simultaneous trials to obtain the right result by chance. He pointed out that there was simply no room for such an event to take place since the earth's surface is only 5×10^{18} cm^2. He suggested that the properties of the molecule had been a decisive factor in sorting out configurations, thus making some more probable than others.

44

It Took the Last 20 Years to Demonstrate that What Were Considered Chromosome Accidents Were Ordered Events

In chromosome and cell research, every phenomenon which initially was thought to be a product of randomness, has turned out to be highly ordered, once it was investigated at the molecular level. The journey took over 100 years and it is in the last 20 years that the verdict of Laplace turned out to be correct. Randomness was a product of the ignorance of molecular mechanisms.

The transmission of hereditary traits, mutation events, exchanges between chromosome segments, rearrangements occurring within and between chromosomes, as well as a long array of genetic phenomena, have turned out to be guided by molecular events that ensure order at the chromosome level.

This order, in the chromosome and the cell, is mainly orchestrated by a series of RNAs and proteins that intervene at the right moment and right place.

45

The Transmission of Hereditary Traits — From Confusion to the Ability to Predict

Crosses between plants, as well as between animals, resulted in progenies whose relationships could not be understood. Their characteristics were considered to be inherited in a random fashion. G. Mendel in 1857 chose simple traits, such as different seed colours in peas, and recognized that their transmission followed order. One could predict the type of progeny that arose. Exactly what each individual would look like could not be asserted, but the proportions in which the different traits would appear in the next generation could be secured.

Following the publication of the "Origin of Species" two years after Mendel's work (1859), no one was interested in order in biology and his work went into oblivion.

In the mean time embryology had developed into an independent science, showing that there was an impressive order in embryonic development which was common to species as diverse as fish and humans. This knowledge redirected biology to a renewed search for order in inheritance. A better proof could not be obtained of this new approach. The problem became so pressing that three different scientists, in three different countries, and at the same time devised experiments to test the consequences of crosses in plants. The year was 1900. They were H. De Vries in

Holland, C. Correns in Germany and E. Tschermak in Austria. The three biologists rediscovered the rules of inheritance. Where before there was disorder prediction became possible and genetics emerged as an independent science.

Commenting on the importance of the discovery of Mendel's laws, the American geneticists Sinnott and Dunn (1939) stated: "The chief purpose of scientific investigation is to extend existing knowledge of these laws over a wider and wider field until all facts which now seem confused and irregular shall take their places in an orderly system".

46

Mutation Has Been the Main Example of a Random Event

A mutation is a sudden change in an organism's trait which becomes stable and is transmitted to the progeny. Like other new genetic phenomena it was attributed to random events in the hereditary material. This interpretation was reinforced when mutations were induced artificially with the help of X-rays and later with chemicals. These agents had a violent effect on chromosomes breaking them into pieces and creating numerous unexpected mutations. Radiation and mustard gas (used in war) were so devastating in their effects that it was like putting the chromosome in front of a death squad. Errors and accidents were considered the source of mutation. As the molecular knowledge of DNA increased it was found that mutations could be of many types and mutation became defined as a "change in a nucleotide sequence of a short region of a genome" (Brown, 1999). The genome is the sum of all the genetic information of an organism.

47

How the Prevailing Fashion Led to Fashionable Results, or How Random Mutations Turned Out to be Non-Random

Every scientist is a prisoner of the ideas that dominate during his lifetime. When he conceives the details of an experiment his mind is canalized by the technology available and by the general current of thought. He, or she, can hardly do otherwise. They are intellectually chained to the interpretations that prevail in their science at the time.

The race for creating an atom bomb started with the beginning of World War II (1939–1945). But, in case the bomb could not be built in time other weapons, such as the biological ones, had to be prepared. It was then that knowledge of the genetics of bacteria became imperative. This frame of mind led to the first experiments in this area.

In 1943 Luria and Delbrück (who later received Nobel prizes) planned an experiment to test the occurrence of mutation in bacteria. During World War II, and later on, to think in terms of randomness was in fashion. They submitted cultures of the bacterium *E. coli* to the action of the bacterial virus T1. Most cells died but some survived the infection. Mutations that were resistant to the virus were considered to occur randomly at various times. Their conclusion was cited, time and again, as the proof of mutation's randomness.

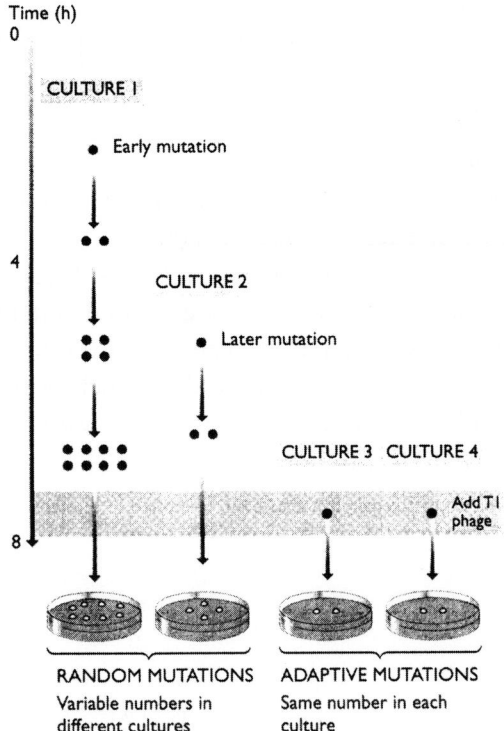

Fig. 3.5   During world war II (1939–1945) the need to develop bacteriological warfare led to an intense investigation of the genetic properties of microorganisms.

In 1943 Luria and Delbrück grew a series of cultures of the bacterium *E. coli* in different flasks and then added the T1 bacteriophage (a virus that attacks bacteria). Most bacteria were killed by the phage, but a few T1-resistant mutants were able to survive. They concluded that early and later random mutations had occurred (cultures 1 and 2). This experiment was later repeated by Cairns *et al.* (1988) who showed that when lactose was the only sugar available, as a source of energy to the bacteria, these produced mutants, with specific DNA changes, in a high number. This allowed the bacteria to adapt to the new medium. The adaptation was characterized by the production of the same number of mutants in each culture (cultures 3 and 4).

One had to await until 1988 for the experiment to be repeated, but now looking for the possibility of another type of response from the same bacterium (Cairns *et al.*, 1988). A strain of *E. coli* was found which could not use the sugar lactose as an energy source. When the cells were plated on a medium containing lactose as the only sugar, the number of mutants that arose enabling the bacteria to use the sugar, became suddenly significantly high. The cells responded to the adverse medium by creating mutations in a direction that allowed them to survive. Mutation was not random (Fig. 3.5). Numerous attempts have been made to discover flaws in this experiment, wihout success and similar results have been obtained with other bacteria (Brown, 2007).

The prevailing fashion in 1943, led to produce a result that, although not falsified, or incorrect, was carried out under conditions difficult to access in detail. In reality the experiment was quite preliminary. However, since it allowed a conclusion that agreed with the general lore it was accepted at its face value.

48

Rearrangements that Were Random Events Turned Out to be Directed by Mobile Elements

The chromosome changes its structure permanently. By using its "magical tricks" it is able: (1) To be broken and immediately resealed. (2) To have its segments inverted but continuing to function with equal efficiency. (3) To exchange DNA sequences with other chromosomes. (4) To delete or add extra copies of its genetic material. The chromosome has been, and is being restructured, at every moment of its cellular existence. Surprisingly, having been so extensively transformed for over 2 billion years, it has maintained its organization.

Human chromosomes are not exceptions to these events. All known types of rearrangements also occur in the human complement. We have existed, as a species, for over 1 million years, and every human egg goes through thousands of cell divisions during the production of an adult. Yet human chromosomes have not disintegrated.

From the beginning geneticists tended to consider all rearrangements as random events. As the molecular knowledge increased, the DNA sequences mainly responsible for this process were identified. These have been named transposable elements or transposons. They are mobile and can migrate to other regions of the same chromosome or to other chromosomes. They consist of a region of insertion into DNA which is flanked by duplicate sequences, of which the DNA bases are known.

Human DNA transposons have terminal inverted repeats and encode the enzyme transposase which regulates transposition. In humans all classes of transposons occupy 45% of the genome, and in the mouse this figure is 37.5% (Dennis, 2002). All mammalian transposable elements characterized to date, appear to be nonrandomly distributed in the genome (Wichman *et al.*, 1992).

Transposons have been called "jumping genes" because about 100 are actively jumping in humans and about 3,000 are thought to be on the move in mice chromosomes (Dennis, 2002).

49

The Repeat Sequences of the Human Genome Are, Instead of Being "Junk", a Treasure Trove of Information

The sequencing of the DNA bases of the human genome by the International Human Genome Sequencing Consortium (2001) has unveiled much unexpected genetic information.

In humans, sequences coding for proteins and RNAs, comprise less than 5% of the genome, whereas repeat sequences account for at least 50% and probably much more.

The repeats fall into five classes. (1) Transposon derived repeats. (2) Inactive copies of cellular genes (pseudogenes). (3) Simple sequence repeats. (4) Segmental duplications. (5) Blocks of tandemly repeated sequences such as centromeres and telomeres.

As the authors of the report point out "Repeats are often described as "junk" and dismissed as uninteresting. However, they actually represent an extraordinary trove of information about biological processes." The repeats are seen, at present, as a rich "palaeontological record" holding crucial clues about the earlier evolution of the chromosome. They are considered to have been, and continue to be, active agents that reshape the chromosomes by causing rearrangements, creating new genes, modifying existing ones and modulating the overall DNA base content.

50

Genes with Similar Functions Could Not be Located Nearby — Random Mutations and Rearrangements Would Disrupt Any Possible Order

When immunologists started to publish data, showing that genes with similar functions were located close to each other, geneticists dismissed such results as incorrect. The occurrence of random mutations and random rearrangements would not permit the establishment of a functional order along the chromosome.

Since then the evidence accumulated has disclosed that, in all types of organisms, genes with similar functions tend to be clustered, building a functional package. Moreover, the assembly has been preserved throughout evolution, the gene cluster being transmitted with little or no alteration.

Gene clustering related to function is already present in viruses and it extends from bacteria to humans. This represents an enormous stretch of evolutionary history. The functions that the clustered genes perform are most diverse. They include the production of: (1) chlorophyll in maize, barley and peas, (2) proteins in the fly *Drosophila* and the mouse, (3) ribosomal RNA in the frog *Xenopus*, (4) eye pigment in humans.

In mice and humans the best known example is the major histocompatibility gene complex that occupies a large DNA segment of 3,500 kilobase pairs long on chromosome 6 in humans and

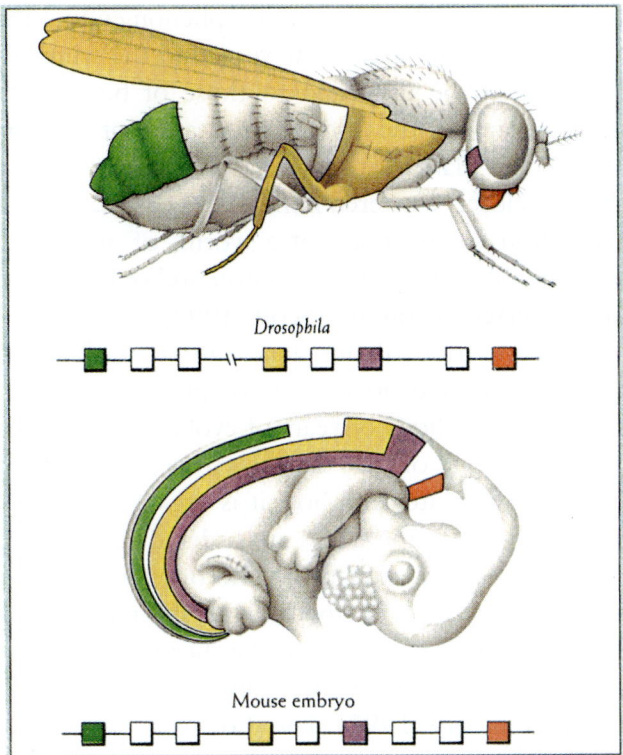

Fig. 3.6   The position of the genes along the DNA sequence follows the position of the parts of the body on which their action is exerted and has not changed from insects to humans.

Comparison of the "homeobox" genes in the fly *Drosophila* and mouse embryo, showing the parts of the body that they influence. The different genes that build the "homeobox" are represented as coloured squares located along the DNA sequence. The genes have maintained the same internal order throughout millions of years of evolution. Moreover, the sequence of their effect on the construction of the adult organism has not been altered during this period.

chromosome 17 in mice. This complex controls many activities of immune cells.

The clustering of the genes producing ribosomal RNA is characterized by hundreds of genes being located close to each other in the chromosome that builds the nucleolus (450 copies in *Xenopus*).

One of the most striking cases of this phenomenon is the home-obox gene family. Until only a few years ago the wings of insects were considered to be unrelated to those of birds and no one would dare to compare the body segmentation of a worm, or the body pattern of a plant, with that of a human. At present it is known that the closely clustered homeotic genes are responsible for the segmentation of the body of a worm, of an insect, of the sequence of the different parts of a flower and of the segmentation of the human vertebral column (Scott, 1992).

But the organization of this gene cluster involved other features: Within the complex its various components have maintained their order during millions of years of evolution. As Alberts *et al.* (1994) point out, one of the "deep mysteries" of the homeobox domain is how it operates and how it is organized. As they state: "the sequence in which the genes are ordered along the chromosome corresponds almost exactly to the order in which they are expressed along the axis of the body". But still more remarkable is that this feature of the gene complex has been highly conserved in the course of evolution being the same in flies and in humans (Warmus and Weinberg, 1993) (Fig. 3.6).

51

The Gene Turns Out to Consist of a Highly Ordered Procession of DNA Stations Locked by Well Defined Starting and Finishing Sites

It is difficult to think of a biological concept that has evolved so rapidly and changed so much as the gene concept, but this is a salutary sign because it derives from the vigorous development of genetics and molecular biology during the last decades.

Carlson (1966) who wrote a critical history of the gene concept has summarized this evolution. "The gene has been considered to be an undefined unit, a unit-character, a unit factor, a factor, an abstract point on a recombination map, a three-dimensional segment of an anaphase chromosome, a linear segment of an interphase chromosome, a sac of genomeres, a series of linear subgenes, a spherical unit defined by target theory, a dynamic functional quantity of one specific unit, a pseudoallele, a specific chromosome segment subject to position effect, a rearrangement within a continuous chromosome molecule, a cistron within which fine structure can be demonstrated, and a linear segment of nucleic acid specifying a structural or a regulatory product".

At present, the sequencing of the bases covering many different types of genes, from bacteria to humans, has led to an accurate picture of its internal organization. An example is the DNA sequence for beta-globin which leads to the production of the protein hemoglobin.

A close molecular look discloses a procession of defined DNA stations. These are ordered and locked within a frame of starting and finishing signals which are themselves composed of small groups of DNA bases. The whole starts with a promoter region where RNA polymerase binds initiating the production of an RNA molecule from the DNA code (Fig. 3.7).

Transcription is the mechanism by which the DNA code is transformed into an RNA message and translation is the process by which this message is converted into a protein. The promoter is followed by a transcription initiation site, a leader sequence (or untranslated region) and a translation initiation codon (i.e. a sequence of only three nucleotides). The regions which will participate in the message for the protein are exon 1, intron 1, exon 2, intron 2 and exon 3. However, only the three exons will be present in the final message used to build the protein. Locking effectively the gene function is the translation terminator codon and a transcribed but not translated region (this includes the poly(A) tail necessary to confer stability on the messenger RNA). Transcription continues beyond the untranslated region for about 1,000 nucleotides until it is definitely capped by the transcription terminator sequence (Fig. 3.7).

What a diversification of functions! But this is not enough by itself. From the DNA sequences of the gene to the production of the final hemoglobin four independent processes are obligatory and these are equally rigidly ordered. The first is the transcription of the long nuclear RNA which is then spliced into a shorter messenger RNA when introns 1 and 2 are removed. It is this last RNA that is translated into the sequence of amino acids that form the beta-globin protein, and it is by posttranslational modification that this protein assumes a totally different shape becoming hemoglobin (Fig. 3.7).

At this last stage an unexpected event occurs which is not directly related to the DNA sequence of the gene. The beta-globin snatches 4 iron atoms from the interior of the cell in the form of heme complexes, and incorporates them in a specific crevice that

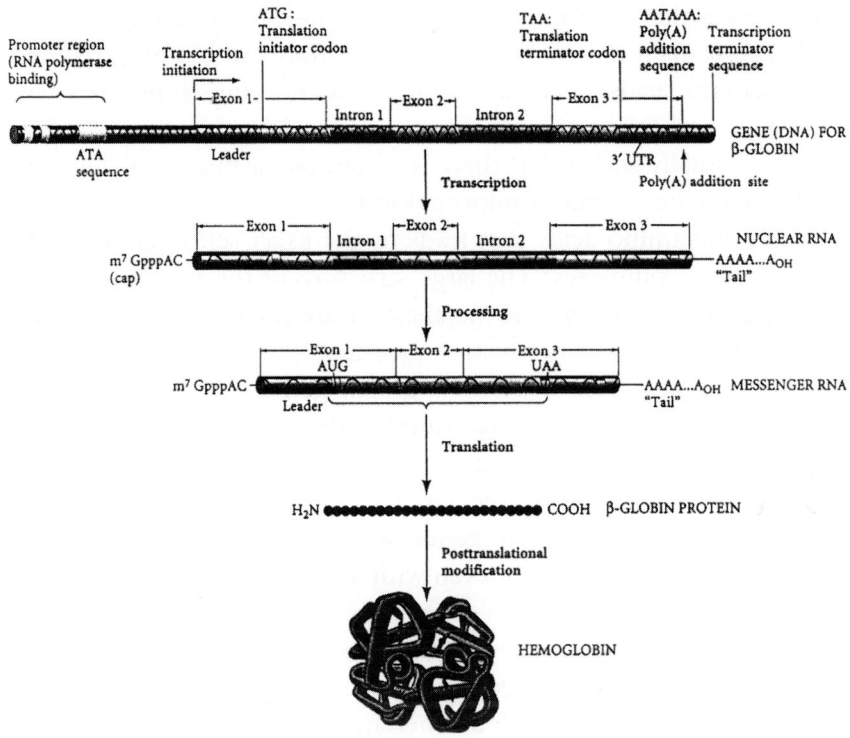

Fig. 3.7   The gene consists of DNA stations that follow each other in a procession like a holiday parade.

The DNA of the gene for beta-globin, like that of other genes, is a most ordered structure. It consists of a series of base sequences whose arrangement cannot be altered and without which it will not function. A promoter region is followed by a transcription initiation site and a translation initiator codon (ATG). Exon 1 is followed by: intron 1, exon 2, intron 2 and exon 3. The termination is carried out by a translation terminator codon (TAA) followed by 3'UTR, which in turn is followed by the Poly(A) addition sequence (AATAAA) and the Poly(A) addition site. Finally the transcription terminator sequence closes the very end of the gene. The ensuing transcription, processing, translation and posttranslational modification, are also subject to an equally impressive order, otherwise the same final molecule (hemoglobin) would not be the obligatory product. The gene (DNA) for beta-globin is transcribed into the large nuclear RNA (consisting of exons and introns). This RNA is then processed into a smaller messenger RNA by discarding the introns. The translation of the messenger RNA results into the formation of a beta-globin protein which consists of a long chain of amino acids. By a posttranslational modification the chain changes its shape, acquires four atoms of iron, and becomes converted into the protein hemoglobin which is the oxygen-carrying pigment of red blood cells.

has always the same molecular pattern, and which confers to the protein its main function, i.e., the iron capacity of binding oxygen. The amino acids of beta-globin know nothing about oxygen uptake.

It is usually stated that there is a gene for hemoglobin. But no DNA sequence contains information for iron. It is the configuration of the amino acids that leads to the exact self-assembly with these metal complexes. The large structure of the macromolecule is like a big cage where the metal atoms are trapped and located in precise sites where they function properly.

The organization of the beta-globin gene, just described, was chosen because it is one of the simplest types of genes, having only two introns. But there are genes built by a whole array of them. The ovalbumin gene in the chicken has 7 introns (Chambon, 1981). There are human genes with 13, 17 and 25 introns (Gitschier *et al.*, 1984) and even with 117 introns, as is the case in the human type VII collagen gene (Christiano *et al.*, 1994). A question then arises. By what mechanism has the gene been able to add over 100 introns, which separate as many exons, being able at the same time to maintain such a coherence that the final product is a perfectly functional protein? We are far from knowing how DNA evolves, in chemical terms, increasing or decreasing its length, yet maintaining an ordered function.

52

The Gene is Never Alone

The DNA of the gene is hopeless without the proteins that bind to it. Without them its message is a black box of no value. The RNA polymerase is the prime motor of the transcription, but it is assisted by at least six mid-wife proteins, which help in the birth of the RNA from the DNA.

Other proteins are located at great distances from the gene, as far away as 50,000 bases. These come from DNA sequences present in the same or other chromosomes. They have been called enhancers, and affect other gene promoters controlling the efficiency and rate of RNA transcription.

Enhancers have turned out to be critical for the regulation of normal development because most genes require enhancers for their activity.

The antithetical nature of the gene and of chromosome behavior is revealed at every level. As there are enhancers which are necessary to promote gene activity, there are also negative enhancers that silence genes. There are even "super-enhancers" that open up several genes at a time. Their action is polarized. The genes which are closest to them are transcribed first and those that are further away are affected later. The distance at which the genes are located determines the order of action (Dillon *et al.*, 1997).

53

The Periodic Packaging of DNA Along Chromosomes Has Turned Out to be Predictable

At present, so much information is accumulating on chromosome organization, that even its simplest structural components turn out to be ordered.

The nucleosome is the fundamental repeating unit of the chromosome material. It consists of 160–240 base pairs of DNA (depending on the species and tissue) wrapped around a core of proteins called histones. In higher organisms there are between 50,000 and 500 million nucleosomes per cell, which are easily visible in the electron microscope.

The position of the nucleosomes along the chromosome has functional importance because they occlude the DNA from interacting with most DNA binding proteins. What is new is that they have turned out to occur periodically along the chromosome and recently it has been shown that their location is so regular that it can be predicted, disclosing that they are not formed randomly along the DNA (Segal *et al.*, 2006).

5 4

In Cell Division the Proper Movement of Chromosomes is Maintained by Correction of Improper Attachment to the Moving Apparatus

The maintenance of order throughout the untold number of cell divisions, is not the product of accidents, but is the result of inherent mechanisms that repair errors. Not only the movement of molecules, and of several cell organelles, is organized, but the movement of chromosomes within the cell is also under control.

If the chromosomes move in an improper way during cell division, the cell resorts to mechanisms that correct this disturbance right away. For accurate segregation of chromosomes, microtubule fibers must attach to centromeres, which are directed to opposite poles of the mitotic apparatus. The enzyme Aurora kinase ensures that the chromosome attaches properly to the spindle fibers. If this enzyme is inhibited, chromosomes fail to orient in the correct way (Lampson *et al.*, 2004). Hence, the cell produces specific enzymes that ensure that chromosomes move in an ordered way (Fig. 3.8).

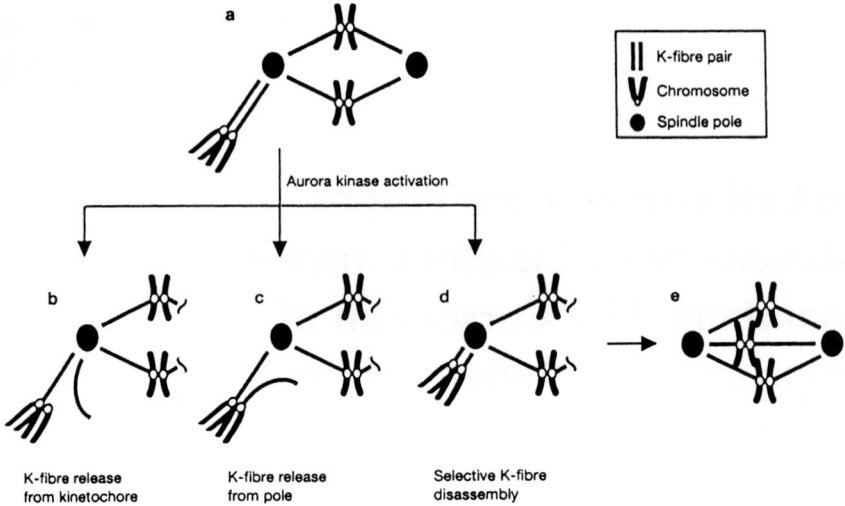

Fig. 3.8   Mechanisms which correct chromosome mal-orientations during cell division.

Each chromosome becomes divided into two chromatids which will move in opposite directions to daughter cells. These chromatids may be prone to errors in spindle attachment during their separation at cell division. Activation of the enzyme *Aurora* kinase is considered to be responsible for ensuring accurate segregation. If the two chromatids happen to be mal-oriented being connected to the same pole by kinetochore or centromere (K) fibres, these are disrupted by *Aurora* kinase activation which ensures a stepwise perfect orientation. (a) Both chromatids happened to be erroneously attached to the same pole by fibres. (b) K-fibre is released from the centromere. (c) K-fibre is released from pole. (d) Selective K-fibre disassembly. (e) Finally the chromatids are aligned with the others ensuring a correct separation.

55

Cells Sense and Stop Uncontrolled Divisions Released by Cancer Stimuli. Moreover, RNAs Are Able to Silence Genes

Human cancer could be much more frequent than it actually happens to be. Since birth, the renewal of the cell population in our body is prone to errors that would lead to a plethora of cancers. In humans, as many as 10^{11} cells die in each adult, each day, and are replaced by other cells (Gilbert, 2000). What an opportunity for total confusion.

One of the reasons why this does not happen "is that normal cells can somehow perceive and arrest aberrant cycles of cell division that are triggered by cancer-promoting (oncogenic) stimuli" (Venkitaraman, 2005).

The mechanism involved is the following. Early cancer lesions result in aberrations in DNA replication. This event triggers the activation of a cellular DNA-damage response (DDR) which arrests cell proliferation or causes cell death inhibiting cancer development (Gorgoulis *et al.*, 2005; Bartkova *et al.*, 2005) (Fig. 3.9).

Recent RNA research has furnished new clues to gene silencing mechanisms. RNAs are usually single-stranded but double-stranded RNAs have turned out to be the key to trigger silencing. It turned out that the double-stranded RNA was a more potent silencer than either strand alone, a phenomenon called RNA interference. In

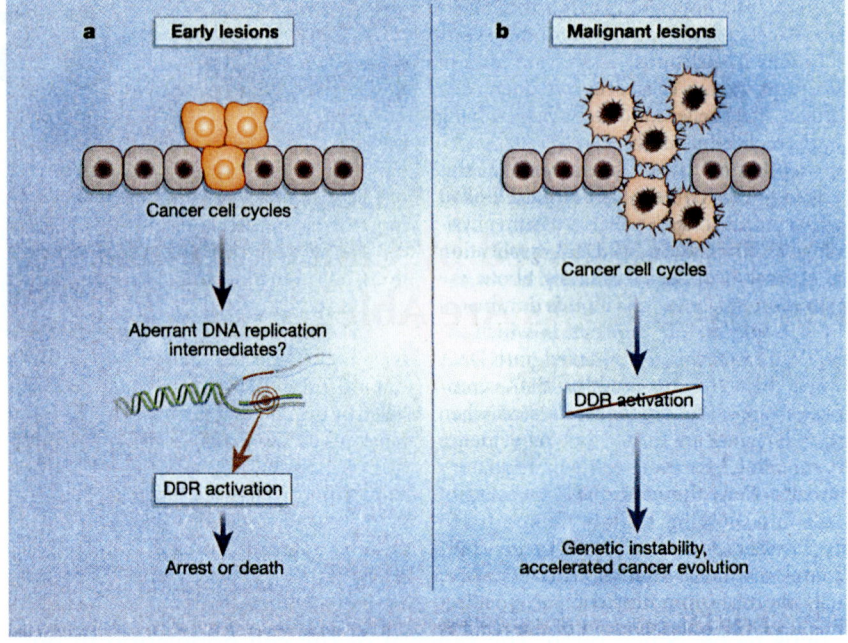

Fig. 3.9 Sensing and stopping wayward cell divisions.

A. In early cancerous lesions, cell-divisions driven by oncogenic stimuli, result in aberrations in DNA replication. This triggers the activation of the cellular DNA-damage response (DDR). The mechanism arrests cell proliferation or causes cell death.

B. On the contrary the progression of malignancy may be accompanied by DDR inactivation which would accelerate cancer evolution and tumor growth.

addition, small interfering RNAs have been found to be part of the pathway. These RNAs act by being complementary to the target of silencing. The small RNAs are transferred to the Argonaute associated protein complexes which lead to messenger RNA degradation, cleavage and other silencing events (Denli and Hannon, 2003) (Fig. 3.10).

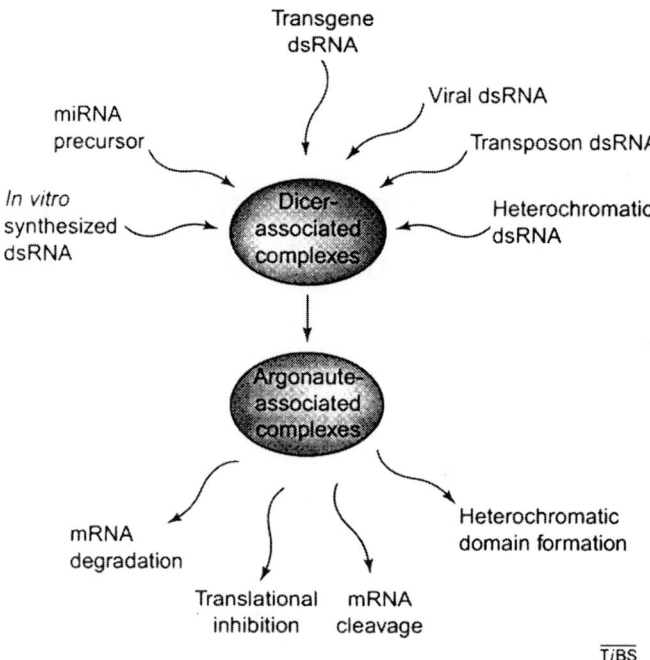

Fig. 3.10   Minute cell RNAs and interfering RNAs are involved in silencing genes.

Diagram showing multiple inputs to the RNA interference machinery, which lead to distinct types of gene silencing events. Double-stranded RNA (dsRNA) from a variety of sources gets processed by Dicer-associated complexes (which produces small interfering RNAs, siRNAs). This guides Argonaute-associated complexes for downstream processes. Dicer and Argonaute family proteins are main components of the RNA-induced silencing pathways, which lead to messenger RNA degradation, mRNA cleavage and other silencing events.

ds = double-stranded RNA,
mi = micro or small RNA,
m = messenger RNA,
si = small interfering RNA.

56

Prevention of Failing of Chromosome Pairing and of Recombination

Other enzyme mechanisms are at work that prevent other types of disturbances.

Meiosis, consists of a series of divisions in which the chromosome number is reduced to half in the ensuing sexual cells. Two obligatory features of this event are chromosome pairing and recombination. The chromosomes of the father and those of the mother come together and build pairs. During this embrace they exchange segments of DNA, a regular process called recombination, which results in novel gene combinations that turn out in the progeny. Hence there is an alteration in the DNA sequence along the chromosome which is an obligatory natural event.

For many years the pairing and the exchanges of DNA sequences were considered to be due to mechanical processes. But at present it is known that if the usual pairing and subsequent recombination of the chromosomes fail to happen there are proteins that intervene at once to prevent such a mishap.

This "safety valve" has been called "The pachytene checkpoint" because it occurs at the beginning of meiosis at a stage called pachytene. What really matters is that this control mechanism prevents chromosome irregularities that would lead to the production of sexual cells with defective chromosome numbers. The "check point" requires a subset of proteins which are specific of this stage of meiosis. As the authors state these proteins "enforce the proper order of the cell-cycle events" (Roeder and Bailis, 2000).

Moreover, as Lam *et al.* (2006) point out "the fidelity of genetic inheritance is orchestrated by a number of evolutionary conserved mechanisms". These are: (1) Chromosome arm cohesion, which holds newly replicated sister chromatids in close proximity and (2) Chromosome condensation which is essential for accurate chromosome transmission during cell division. Remarkable is that these two mechanisms were conserved from yeast to humans.

These observations and experiments demonstrate the presence of hidden mechanisms that the chromosome and the cell carry in their molecular arsenal and which they pick up and use when novel circumstances arise. What a formidable ability to maintain an independence of behavior and what a capacity to circumvent deviations from the original order.

57

Brownian Motion — The Trap of the Physicist and Biologist

The young Scottish doctor Robert Brown (1733–1858), looking through his microscope, examined a drop of liquid on which were suspended pollen grains. He observed that these moved continuously and in every direction, each following a zig-zag path independent of the other grains. Later, the particles which are suspended on a cell cytoplasm were seen to behave in the same way. This strange and disordered movement became called Brownian motion. It was found to be spontaneous and never stopped (Amaldi, 1966).

Physicists were intrigued by this phenomenon. It influenced the random concepts that have dominated elementary particle physics. It was found that the disorderly movement of particles in suspension, is due to the collisions which these undergo with the molecules of the liquid, as demonstrated by the French physicist J. Perrin (1870–1942).

Pauling and Zuckerkandl (1972) were among the first to expose the simplistic interpretation of Brownian movement. They made their point clearly: "It was shown by Einstein and by Smoluchowski half a century ago that by the methods of statistical mechanics, with use of the completely deterministic equations of the classical laws of motion, the behavior of the particle, in a statistical sense, can be completely predicted. The exact motion of the particle — the statement of its positions and velocities as a function of the time — cannot, however, be predicted, because of the lack of information by the observer of the values, at one instant of time, of the positions and momenta of all of the molecules in the fluid (as well as of the molecules in the walls of the system, with which the fluid interacts). It is said to be chance that causes the particle to move along its unpredictable path, in the way in which it is, in one series of observations, observed to move; but it is believed that this "chance" is simply the necessary result of the molecular collisions, called chance because of lack of knowledge of the observer." In later years physicists abandoned their interest in this movement, as they started to search for order in the organization of matter. They now look for "The Origins of Order" as described in the book written by Kauffman (1993).

58

The Cell Was Seen as a Pea Soup, But Now Most of Its Molecules Are Known to Have an Address

On equal grounds biologists were originally most impressed by the Brownian motion which they could easily observe in living cells. This was the trap that contributed to establish the idea that the chemical components of a cell were moving around in a most disordered way. Gene products would find each other by bumping into one another due to random collisions. Randomness could not be easily denied since it could be seen with one's own eyes but it turned out that the cell circumvented this situation by giving its molecules an exact address.

Until recently, the cell was looked like a "pea soup" in which molecules, and its minute cell organelles, met each other accidentally.

This picture turned out to be incorrect. The electron microscope disclosed that most cell organelles are anchored on different membranes and have well defined positions within the cell. They may move, such as the chloroplasts, relocating themselves according to the strength of light, but this movement also occurs in an ordered manner, since they return to their initial location once the light conditions change. Chromosomes are also orderly located inside the nucleus being attached to the nuclear membrane at specific sites (Fig. 3.11).

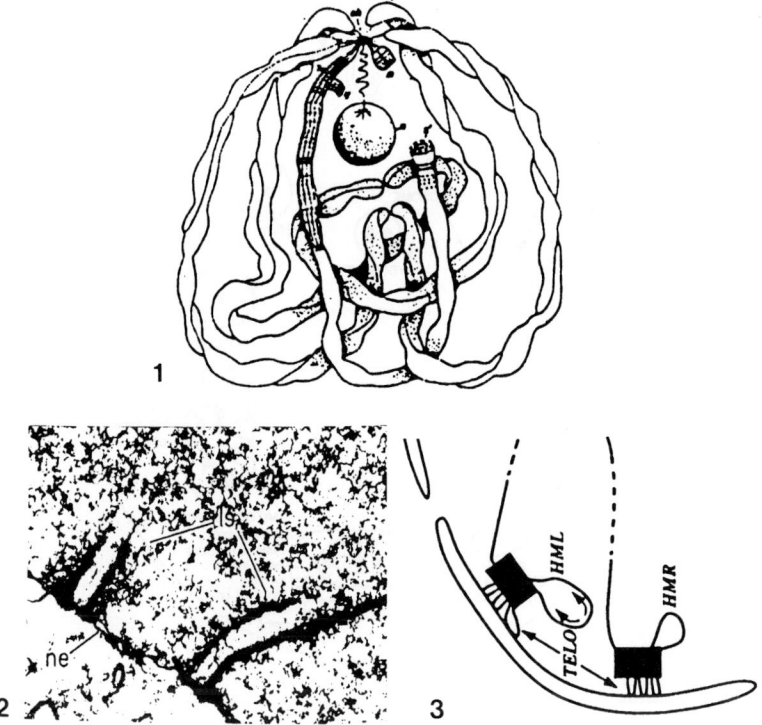

Fig. 3.11 Most cell components are anchored on different structures.

1. The nucleolus, in the salivary gland chromosomes of the fly *Drosophila funebris*, is suspended from the site where the chromosomes are attached to each other.
2. Electronmicrograph showing the termination at the nuclear envelope (ne) of two chromosomes in a rat spermatocyte.
3. The telomeres of the chromosomes of the yeast *S. cerevisiae* associated at the nuclear envelope.

The macromolecules furnish the clearest example of orderly behavior. They have an address label on their molecular frame. The targeting signals that many proteins contain are written on short sequences of their amino acids. With this address, the proteins are targeted, with high specificity and efficiency, to a variety of locations within and outside the cell. Other proteins get their labels after they have entered the Golgi apparatus, an organelle consisting of a compact system of cell membranes.

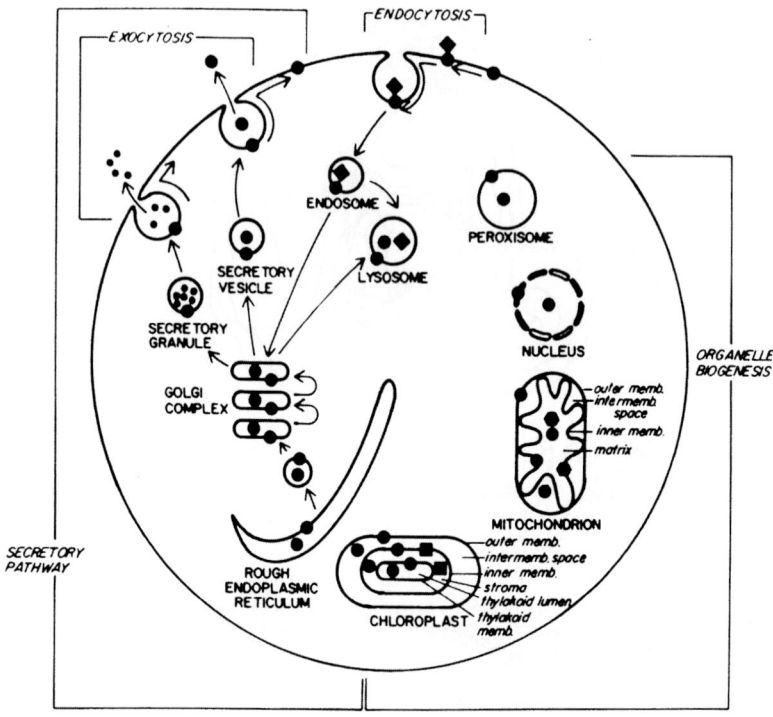

Fig. 3.12 Molecules in the cell have an address.

Diagram of the cell of a higher organism, showing the sites to which targeted proteins are directed. Nuclear-encoded proteins are directed to a variety of sites. Proteins synthetized by other cells are imported via endocytosis. The Golgi complex is the main organelle that furnishes the proteins with an address.

Most "Proteins have signals which tell them where they are to go and how they are to fold and interact with other cell components once they reach their destination" (Pugsley, 1989). We are far removed from the chaos of Brownian motion (Fig. 3.12).

59

"The Genetic Code is Certainly Not Random"

The genetic code is the molecular system by which the information present in DNA is correctly transformed into that of proteins. Groups of three nucleotides in DNA specify the amino acid sequence of the proteins.

As Bacher *et al.* (2004) point out: "Although the genetic code was discovered some 40 — odd years ago, there are still numerous questions as to how it arose and evolved". Initially it was considered by F. Crick a "frozen accident", but soon it was found to have evolved leading to several examples of altered forms, such as in mitochondria. Studying the relationships between amino acids and nucleotides Lacey *et al.* (1993) concluded that "the genetic coding system is certainly not random; rather it appears to be based on character relationships between amino acids and their anticodon nucleotides". The same view was recently expressed by Itzkovitz and Alon (2007) who stated, on the same grounds, that: "The genetic code has been shown to be nonrandom".

60

Noise is Disorder — Music is Order and Unity

To understand what is science as well as art, it is necessary to become aware of the difference between noise and music. Why do we avoid one and choose the other?

Kaufmann (1947) compares and describes the physical difference between noise and music. "The roar of subway trains are not to be confounded with music, for their waves, if photographed, would look jagged as a streak of lightning". On the other hand "When a musical sound is uttered, its waves, are fully regular and properly timed".

And she adds: "The method of varying the duration of tones is not haphazard, but an orderly, mathematical process". Melody: "By definition, it is a succession of musical sounds which have been organized in some kind of coherent shape or pattern". These statements are a key not only to the understanding of music but of science.

Only when there is order and coherence in the explanation of a physical or chemical phenomenon does it become possible to define it in a way that leads to valid predictions. A solar eclipse, or the synthesis of DNA in the cell, are only predictable by using well-defined equations and well-defined molecular interactions, respectively.

61

The Distinction between Genetic Noise and Genetic Music

In its permanent molecular activity, and the reshaping of its structure, the chromosome is prone to faulty solutions in its molecular edifice. These take the form of base deletions, base substitutions, accidental rearrangements and other errors that do not agree with its initial construction. These lead to defective mutations of various kinds and may be called genetic noise.

But there is another type of transformation occurring in the chromosome. Thanks to its inherent mechanisms of surveillance and maintenance of order, these transformations result in what could be called genetic music. As mentioned in the preceding pages, the chromosome escapes from randomness by using various mechanisms: (1) Mutations turn out to be ordered events. (2) Rearrangements are directed by mobile elements. (3) Closely clustered genes with similar functions have been preserved from worms to humans. (4) Genes were found to consist of a procession of DNA stations with starting and finishing signals. (5) If improper movement of chromosomes ensues, the abnormality is corrected by specific enzymes. (6) Cells sense and stop uncontrolled divisions that would otherwise lead to abnormal development. (7) When chromosome pairing and recombination fail to happen the proper solution is enforced by specific proteins. (8) Most cell macromolecules carry an address written on their molecular framework.

As Brown (1999) has stated "randomness does not apply to all components of the non-coding DNA. In particular, transposable elements and introns have interesting evolutionary histories".

In evolution what counted was the genetic music that maintained harmony. This resulted in the emergence of novel organisms that at the same time had a coherent organization. On the contrary, the genetic noise, that occurred since the dawn of life, was disposed of as irrelevant and forgotten by the chromosome.

62

"Errors" Are Not of All Possible Kinds

Although not usually mentioned, but most significant, is that when the copying is not an exact replica of the original DNA sequence, not all kinds of alternatives are possible as it would be in a random situation. In the so called "errors", bases are not substituted by all possible kinds of chemicals available in the cell. Only base analogues (which are chemically similar) or only bases that have the same chemical formula but differ solely in the position of one or a few atoms within the molecule, are incorporated during replication. Thus, there are chemical constraints on the alternatives accepted. The "errors" are only of a particular type and should instead be called limited alternatives.

Surveying DNA replication Strachan and Read (2000) conclude that "The frequency of individual base substitutions is nonrandom" and also that "The location of base substitutions in coding DNA is nonrandom". This means that even genetic noise is not of all possible kinds.

References

Part III

Collins Publishers. (1987) *Collins Cobuild English Language Dictionary.* Collins, London, UK.

Webster N. (1976) *Webster's New Twentieth Century Dictionary.* Collins World, USA.

Honour H, Fleming J. (2002) *A World History of Art.* Laurence King Publishing, London, UK.

Bynum WF, *et al.* (1983) *Dictionary of the History of Science.* Macmillan Press, London, UK.

Katz VJ. (1993) *A History of Mathematics, An Introduction.* Harper Collins College Publishers, NY, USA.

Rebière A. (1900) Laplace. In: *Pages Choisies des Savants Modernes,* pp. 125–139. Librairie Vuibert, Paris, France.

Darlington CD. (1953) *The Facts of Life.* George Allen and Unwin, London, UK.

Eden M. (1967) Inadequacies of neo-Darwinian evolution as a scientific theory. In: PS Moorhead & MM Kaplan (eds), *Mathematical Challenges to the Neo-Darwinian Interpretation of Evolution,* pp. 5–12. Symposium No. 5, The Wistar Institute of Anatomy and Biology, The Wistar Institute Press, Philadelphia, USA.

Pauling L, Zuckerkandl E. (1972) Chance in evolution — some philosophical remarks. In: DL Rohlfing & AI Oparin (eds), *Molecular Evolution,* pp. 113–126. Plenum Press, NY, USA.

Mora PT. (1965) The folly of probability. In: SW Fox (ed), *The Origins of Prebiological Systems and of Their Molecular Matrices*, pp. 39–64. Academic Press, NY, USA.

Haldane JBS. (1965) Data needed for a blueprint of the first organism. In: SW Fox (ed), *The Origins of Prebiological Systems*, pp. 11–18. Academic Press, NY, USA.

Sinnott EW, Dunn LC. (1939) *Principles of Genetics*. McGraw-Hill Book Co, NY, USA.

Brown TA. (1999). *Genomes*. Bios Scientific Publishers, Oxford, UK.

Luria SE, Delbrück M. (1943) Mutations of bacteria from virus sensitivity to virus resistance. *Genetics* **28**: 491–511.

Cairns J, *et al.* (1988) The origin of mutants. *Nature* **335**: 142–145.

Brown TA. (2007). *Genomes 3*. Garland Science, NY, USA.

Dennis C. (2002) A forage in the junkyard. *Nature* **420**: 458–459.

Wichman HA. *et al.* (1992). Transposable elements and the evolution of genome organization in mammals. *Genetica* **86**: 287–293.

Dennis C. (2002) A forage in the junkyard. *Nature* **420**: 458–459.

International Human Genome Sequencing Consortium. (2001) Initial sequencing and analysis of the human genome. *Nature* **409**: 860–921.

Scott MP. (1992) Vertebrate homeobox gene nomenclature. *Cell* **71**: 551–553.

Alberts B, *et al.* (1994) *Molecular Biology of the Cell*. Garland Publishing, Inc, NY, USA.

Varmus H, Weinberg RA. (1993) *Genes and the Biology of Cancer*. Scientific American Library, NY, USA.

Carlson EA. (1966) *The Gene: A Critical History*. WB Saunders Co, Philadelphia, USA.

Chambon P. (1981). Split genes. *Sci Am* **244(5)**: 48–59.

Gitschier J, *et al.* (1984) Characterization of the human factor VIII gene. *Nature* **312**: 326–330.

Christiano AM, *et al.* (1994). Structural organization of the human type VII collagen gene (COL7A1), composed of more exons than any previously characterized gene. *Genomics* **21**: 169–179.

Dillon NT, *et al.* (1997) The effect of distance on long-range chromatin interactions. *Mol Cell* **1**: 131–139.

Segal E, *et al.* (2006) A genomic code for nucleosome positioning. *Nature* **442**: 772–778.

Lampson MA, *et al.* (2004) Correcting improper chromosome-spindle attachments during cell division. *Nat Cell Biol* **6**: 232–237.

Gilbert SF. (2000) *Developmental Biology.* Sinauer Associates Publ, Sunderland, MA, USA.

Venkitaraman AR. (2005) Aborting the birth of cancer. *Nature* **434**: 829–830.

Gorgoulis VG, *et al.* (2005) Activation of the DNA damage checkpoint and genomic instability in human precancerous lesions. *Nature* **434**: 907–917.

Bartkova J, *et al.* (2005). DNA damage response as a candidate anti-cancer barrier in early human tumorigenesis. *Nature* **434**: 864–870.

Denli AM, Hannon GJ. (2003) RNAi: an ever-growing puzzle. *Trends Biochem Sci* **28**: 196–202.

Roeder GS, Bailis JM. (2000) The pachytene checkpoint. *TIG* (September 2000) **16**: 395–403.

Lam WW, *et al.* (2006) Condensin is required for chromosome arm cohesion during mitosis. *Genes Dev* **20**: 2973–2984.

Amaldi G. (1966) T*he Nature of Matter.* University of Chicago Press, Chicago, USA.

Pauling L, Zuckerkandl E. (1972) Chance in evolution — some philosophical remarks. In: DL Rohlfing & AI Oparin (eds), *Molecular Evolution*, pp. 113–126. Plenum Press, NY, USA.

Kauffman SA. (1993) *The Origins of Order.* Oxford University Press, NY, USA.

Pugsley AP. (1989). *Protein Targeting.* Academic Press, NY, USA.

Bacher JM, *et al.* (2004). Evolving new genetic codes. *Trends Ecol Evol* **19**: 69–75.

Lacey JC, *et al.* (1993) Couplings of character and of chirality in the origin of the genetic system. *J Mol Evol* **37**: 233–239.

Itzkovitz S, Alon U. (2007) The genetic code is nearly optimal for allowing additional information within protein-coding sequences. *Genome Res* **17**: 405–412.

Kaufmann H. (1947) *The Little Guide to Music Appreciation*. Grosset and Dunlap, NY, USA.

Brown TA. (1999). *Genomes*. Bios Scientific Publishers, Oxford, UK.

Strachan T, Read AP. (2000). *Human Molecular Genetics*. Wiley-Liss, NY, USA.

Sources of Illustrations

Part III

3.1 Studenbrock C, Töpper B. (1999) 1000 *Mästerverk i den Europeiska Bildkonsten* (Swedish translation). Könemann, Germany (Fig. Astrono[men 1668, Louvre, Paris, by Jan Vermeer, page 918. RMN/Jean C).

3.2 Armstrong R. (1968) *The Discoverers.* Ernest Benn Limited, London, UK. (Fig. 28 Lisbon Harbour in the early 16th century, page 66. From Bry's *America*).

3.3 Mettais V. (1997) *Your Visit to the Louvre.* Art Lys, Paris, France (Fig. of painting by G. de la Tour, page 60. RMN/No photographer).

3.4 Spufford P. (2003) *Pengar och Makt* (Swedish translation). Albert Bonniers Förlag, Stockholm, Sweden (Fig. of tapestry, page 35).

3.5 Brown TA. (1999). *Genomes.* Bios Scientific Publishers, Oxford, UK. (Research Briefing 13.1, page 347).

3.6 (1) Varmus H, Weinberg RA. (1993) *Genes and the Biology of Cancer.* Scientific American Library, NY, USA. (Fig. Genes encoding, page 150).

    (2) Scott MP. (1992) Vertebrate homeobox gene nomenclature. *Cell* 71: 551–553 (Fig. 1, page 551).

3.7 Gilbert SF. (2000) *Developmental Biology.* Sinauer Associates, Publ, Sunderland, MA, USA (Fig. 5.3, page 112).

3.8 Lampson MA, *et al.* (2004) Correcting improper chromosome-spindle attachments during cell division. *Nat Cell Biol* 6: 232–237 (Fig. 4, page 236).

3.9   Venkitaraman AR. (2005) Aborting the birth of cancer. *Nature* **434**: 829–830 (Fig. 1, page 829).

3.10   Denli AM, Hannon GJ. (2003) RNAi: an ever-growing puzzle. *Trends Biochem Sci* **28**: 196–202 (Fig. 2, page 200).

3.11   (1)   Frolowa SL. (1935) In: Darlington CD. (1937) *Recent Advances in Cytology.* Churchill, London, UK. (Fig. 59, page 176).

       (2)   Moses MJ, Coleman JR. (1964) In: M Locke (ed), *The Role of Chromosomes in Developoment,* pp. 14. Academic Press, NY, USA. From DuPraw EJ. (1970). *DNA and Chromosomes.* Holt, Rinehart and Winston, NY, USA (Fig. 15.1, page 250).

       (3)   Gilson E, *et al.* (1993). Telomeres and the functional architecture of the nucleus. *Trends Cell Biol* **3**: 128–134 (Fig. 4, page 132).

3.12   Pugsley AP. (1989) *Protein Targeting.* Academic Press, NY, USA (Fig. 1.1, page 2).

Who Cares for Selection

63

Selection is a Political, Not a Scientific Concept

Charles Darwin (1809–1882) was an excellent naturalist but the expanding chemistry and physics of his time never interested him, as attested by his autobiography (F. Darwin, 1929). This is why, when he was faced with the problem of the origin of the species, he did not turn to these disciplines. Instead he used the sociological doctrines that dominated in his country. In the time of Queen Victoria, when England was becoming a main industrial and colonial power, the Malthus doctrine fitted perfectly the interests of the rich class to which he belonged.

In "An Essay on the Principle of Population as it Affects the Future Improvement of Society", published in 1798, Malthus described the perils of an unchecked population growth and advocated the need to suppress the procreation of the lower classes. As Darwin (1859) describes, he borrowed from Malthus (1766–1834) the concepts of selection and of the survival of the strongest. It was the philosopher Herbert Spencer who pointed out to him that he should change the survival of the strongest into that of the fittest. Already Radl (1930) put it in a simple sentence "Darwin merely transferred the prevailing English political ideas and applied them to nature". His work is referred nowadays as "The Origin of Species", but the original title was "The Origin of Species by Means of Natural Selection or the Preservation of Favoured Races in the Struggle for Life".

Fig. 4.1 The social order of "the richest country in the world" symbolized in 1867.

This cartoon depicts the stratification of society with Queen Victoria at the top of "The British Bee Hive".

It could not be more elucidating. As Hobsbawm (1998) put it: "In that 'struggle for existence' which provided the basic metaphor of the economic, political, social and biological thought of the bourgeois world, only the 'fittest' would survive, their fitness certified not only by their survival but by their domination" (Fig. 4.1).

The result is that the interpretation of evolution became not based on the physical agents and on the chemical components of organisms but described on the basis of population relationships that could not be defined (Fig. 4.2).

Such a view of the biological world, based on an uncontrolled survival of organisms, fitted perfectly the ideology of the Victorian

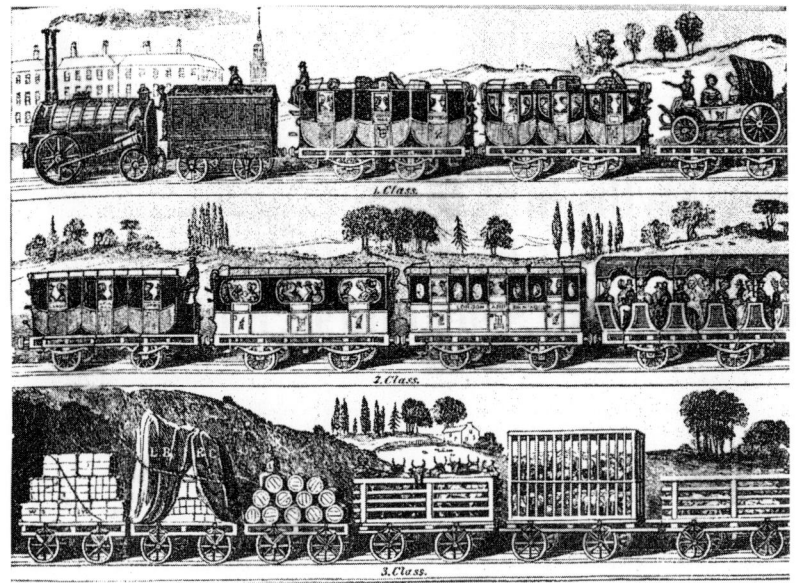

Fig. 4.2   The invention of the railway was a main factor in rapid industrialization and colonial expansion.

During the first half of the 1800s, trains started to carry people and goods throughout Europe and later on in other continents. A picture of the train between London and Birmingham in 1837 loaded with all kinds of merchandise including a vehicle. The train is also filled with people, the result of the increase in population that preoccupied sociologists such as Malthus.

age and it continues to serve the interests and values of the present economic globalization.

This is one of the main reasons why Darwinism is so firmly defended and is so entrenched in our minds.

64

Three Myths in Science:
Phlogiston in Chemistry, Ether
in Physics and Selection in Biology

In science every time an important phenomenon is discovered an explanation is actively sought. What characterizes science is the search for a logical interpretation of a given event. The difficulty is that the agreement between the interpretation and the phenomenon depends on the amount of knowledge available at a given time. The less developed the field of research, the higher the risk that the interpretation just covers the ignorance of the mechanisms involved. But here lies also the strength of the scientific approach, i.e. the need to produce at all times some form of interpretation irrespective of how little it elucidates the mechanism. Such a primary approach is in itself a recognition of the phenomenon and it has the great advantage of obliging later generations to test it and to find out its limitations or erroneous construction. This is because a poor idea can be superseded but the lack of an idea cannot.

There have been three major ideas — one in chemistry (the phlogiston), one in physics (the ether) and one in biology (selection) — that dominated for over a century. All three were not real explanations but initial ways of approaching a new problem. One may then ask why they persisted so long and were so difficult to eradicate. Most people believe that they persisted because they were good explanations. This is not necessarily so; they survived because the phenomena that they 'explained' were important in themselves and because for a long time there was no better substitute. It was the importance of the phenomenon, not the importance of the explanation, that obliged people to defend them at every price. No one would leave a basic phenomenon without some form of logical interpretation (Lima-de-Faria, 1988).

These phenomena were:

(1)   The combustion of substances, which was an event associated with the fire emitted by a burning compound. It dominated the chemistry of the 18th century.
(2)   The propagation of light between celestial bodies, which preoccupied the physicists of the 19th century.
(3)   The transformation of species, a process later called evolution, which emerged as a central theme in 19th century biology.

Three abstractions were created to account for these phenomena. To explain combustion a substance that was called the phlogiston was invented. During combustion the material that burned lost this hypothetical compound.

To explain the transmission of light one invented the ether, a medium destined to function as a support to the propagation of light and that filled the vacuum that was supposed to exist between the planets and the stars.

To explain the evolution of species one invented selection. This represented an abstract interaction between the organism and the environment, which could not be defined in any physico-chemical terms.

What caused the removal from science of some of these concepts and led to them being shelved as historical curiosities?

For over one hundred years most chemists accepted and described as the correct interpretation the 'theory of the phlogiston' developed by the German chemist George Stahl (1660–1734). He was the Darwin of the chemistry of that time. No one, or very few, dared to question his interpretation. In some reactions the phlogiston had to have a positive weight, in others a negative one, and in still others no weight at all. Nothing was left unexplained. It was Antoine Lavoisier (1743–1794), with his experiments on combustion in which he measured the exact weights of the components of the reaction, who showed in an irrefutable way that oxygen was the component involved and that phlogiston simply did not exist.

For many years most physicists accepted and described as the correct interpretation the existence of the ether, which was believed to be a liquid, a solid or to have other properties (Lodge, 1883). The experiments of the physicists A. Michelson (1852–1931) and E. Morley (1838–1923) failed to demonstrate the existence of the ether. But the problem was finally settled when in 1905 Einstein's theory of relativity dispensed with the need for ether (David *et al.*, 2002). The ether, which had been attributed so many properties, did not exist. Whole treatises had been written on it but once the physical nature of light was well defined the ether was abandoned.

Most biologists have accepted, for over one hundred years, the existence of selection as the mechanism of evolution. Selection is equally an ideal solution because it solves every problem for which there is no solution. In some situations selection must be positive, in others negative and in others ought to be neutral. The similarity with the properties of the phlogiston and the ether is evident.

Phlogiston, the ether and selection all three belong to the history of science. They represent periods of endeavor in their respective fields. They were invented because they were the only type of explanation that could be found at the time for a phenomenon which, due to its importance, had to be explained by the means available.

65

Definitions of Selection

Selection has been defined in many ways. A definition is: 'Any process, whether natural or artificial, by which certain organisms or characteristics are permitted or favoured to survive and reproduce in, or as if in, preference to others' (Webster, 1976).

The Russian-American geneticist Dobzhansky (1969) defined it in the following terms. "Selection, in biology, is that process whereby individuals with certain genetic endowments are chosen from a population of animals or plants apparently because they are more suited than others to a given environment. This culling of fitter types, called natural selection, is believed by most biologists to have been the chief impelling and directing agent of organic evolution on earth."

Darwin's original definition of selection (Darwin, 1859) is: "I have called this principle, by which each slight variation, if useful, is preserved, by the term Natural Selection, in order to mark its relation to man's power of selection. But the expression often used by Mr. Herbert Spencer of the Survival of the Fittest is more accurate, and is sometimes equally convenient".

It is to be noted that Darwin calls selection a principle and not a mechanism as assumed by most Darwinists.

In later years selection has been extended from living organisms to molecules and even to atoms (Eigen and Schuster, 1979; Frausto da Silva and Williams, 1994; de Duve, 2005). This is to put the clock back. No book of physics or chemistry has been found to contain in its indexes, or glossaries, the word selection. To substitute the four fundamental forces of physics and their laws by selection, is an expression of the extreme dogmatism that selectionists display at present, an attitude which is foreign to exact science.

66

Selection is Not a Material Component that Can be Measured

Selection cannot be "the mechanism of evolution", because it is not a material component of the universe. Selection cannot be weighed, stored or poured into a vial or measured in specific units. It is only a system of choice and as such cannot be the mechanism. This can only be physico-chemical. Selection is actually the "aspirin" of the biologist. It is a wonderful resource since it actually explains any situation as well as its opposite state.

Molecular biologists who compare the DNA sequences of different genes quite often use the terms 'selection pressure' and 'mutation pressure' to explain the differences or the similarities resulting from base changes (Cavalier-Smith, 1985).

In physics, which is an exact science, there is a phenomenon called atmospheric pressure. This can be measured by an instrument called the barometer and the units in which the pressure of molecules is measured are called kilobars. Molecular biologists, obviously, do not have an instrument to measure their mutation or selection pressure and as a consequence cannot express it in exact units. The use of such a vocabulary tells us how far biology is from being an exact science.

67

The Distinction between Evolution and Darwinism

Evolution is one of the best established phenomena in biology. Comparative anatomy, paleontology, physiology, molecular biology and DNA sequence analysis, all agree in confirming the evolutionary process.

However, there has never been a theory of evolution, in the strict scientific sense of the word. In chemistry and physics, theories are bodies of evidence that allow predictions. It is not possible, at present, to predict what species will emerge after humans, sparrows or lilies. As such, evolution remains a most valuable *working hypothesis*, but is not a theory. The reason is that the mechanism of evolution is not known.

A mechanism is a process totally distinct from the phenomenon itself. In science the understanding of a mechanism usually changes at every generation, as the technology advances. An example from physics is the mechanism responsible for the properties of light. The nature of light switched from being ascribed to particles (Newton), to waves (Maxwell) to particles again (Einstein) and at present to a wave-particle interpretation (DeBroglie).

Genetics is far from being an advanced science. As such one cannot expect, even after the sequencing of the DNA of several organisms, to be in the possession of the mechanism of evolution, since the molecular organization of the chromosome, as a functional unit, is far from being understood (Lima-de-Faria, 2003).

The concept of evolution was initially formulated by Lamarck in 1809 on the basis of the biological data available. Fifty years later (1859) Darwin and Wallace had at their disposal much more information that permitted to give it a more solid basis. But the mechanism remained elusive. Darwin believed in the transmission of acquired characters, since the rules of inheritance were unknown.

68

The Merits and Limitations of Darwinism

Darwinism had initially the great merit of emphasizing the study of the dynamics of natural populations and of giving a great impetus to the study of inheritance leading to the exponential development of genetics. Its empirical approach was of great benefit helping to remove from biology both vitalism and obscurantism. But science, is like any other human activity. It is guided by fashions and fashions turn easily into dogmas. This is why Darwinism has become a dogma.

There is also another reason for Darwinism's entrenched position. Biologists continued to argue. If it is not selection, what is it that directs evolution? And this has been a strong argument that could not be easily pushed aside. Was there an alternative?

An Interpretation of Evolution Based on Physico-Chemical Processes

The sought-after alternative became available when it was shown that evolution did not start with the emergence of the cell. Actually evolution has turned out to be a phenomenon inherent to matter, canalized by the autoevolution of the elementary particles, which have assembled orderly into a limited number of more complex configurations, when quarks and leptons produced protons, neutrons, electrons and other particles. Subsequently, the type of electronic orbitals present in every atom obliged them to have similar properties, which are periodically repeated, as demonstrated by the formation of the Periodic Table of the Elements. As evolution progressed, the self-assembling capacity of atoms led to the formation of a few dominating types of macromolecules, to the production of viruses and cell organelles including the chromosome (see Part VI). This assembling power directed evolution into more diversified channels but at the same time led to restricted numbers of living organisms, which due to their common structural and molecular characteristics do not exceed five kingdoms containing only 32 animal and 9 plant phyla. Evolution at present emerges as a highly ordered, restricted and canalized event and we are far from understanding all the types of mechanisms involved in its origin and further expansion. Since the dawn of matter's evolution a rigid organization prevailed at every level of combination resulting in the intact transmission of the initial properties of the elementary particles as far away as the human species. This is illustrated by the uninterrupted transmission of left-handed and right-handed symmetries, which were already present in the neutrino, and which were transferred intact to the left-handed and right-handed pattern of the human body (Fig. 7.7) (Lima-de-Faria, 1983, 1988, 1997, 2006).

70

How the Chromosome Evades Selection

Among the first to realize the inability of selection to act at the chromosome level were Britten and Kohne (1968). They had demonstrated that hundreds of thousands of copies of DNA sequences had been incorporated into the genomes of higher organisms. When they speculated on their function they were obliged to realize that selection was unable to act on them. The appearance in the genome of many thousands of copies of a gene created new evolutionary situations. They stated: "The dynamics of selection for this set of genes would be fundamentally altered. Owing to the great multiplicity of copies, their selective elimination might be impossible" (Britten and Kohne, 1968).

At present it is known that most DNA sequences occur in the chromosome in multiple copies. For instance, in the toad *Xenopus laevis* there are about 450 copies of the 18S and 28S ribosomal RNA genes in each set of chromososomes (Birnstiel *et al.*, 1966), but there are as many as 9,000 to 24,000 genes for the 5S ribosomal RNA in the same species (Brown *et al.*, 1971). Concerning the human chromosomes, interspersed repeats have been divided into four classes. The so called LINEs and SINEs occur in 850,000 and 1,500 000 respectively. Other elements and the DNA transposons are not less than 450,000 and 300,000 respectively (International Human Genome Sequencing Consortium 2001). This multiplicity of copies was not eliminated by millions of years of selection.

71

A Chaotic Chromosome Could Not Evade Selection but an Organized One Cannot Do Anything Else but Circumvent It

All forms of organization are ways of evading selection. The alternatives on which selection can act become sharply reduced when, from the start, the variations in the molecular and structural organization of the chromosome become confined to a limited number of possibilities imposed by the rigid molecular configurations and reactions.

Each time a molecular relationship is established by stereochemical fitting, involving different types of bonds between two or several molecules, organization is achieved. This new chemical pattern cannot easily be disturbed. It can only proceed along the directions allowed by the chemical reactions established. Once molecular fitting in the chromosome starts to direct the exerting of its properties, such as replication, recombination and conservation (repair), selection is pretty effectively evaded, since these processes represent very rigid canalizations. DNA replication, recombination and repair are not only of universal occurrence but follow the same atomic steps in bacteria and humans. Thus, nothing could disturb their molecular canalization throughout millions of years.

When one considers the structural features of the chromosome the same reasoning applies. The formation of centromeres, telomeres, nucleolus organizers and other specialized structures, emerged as a product of its chemical organization. Each locked the chromosome into a given channel of behavior and as such limited the possibilities of the action of selection. Order at every level of chromosome organization endowed it with a very high degree of self-control. The significance of selection as a 'guiding force' appears bleak. All forms of order represent a hindrance to selection since they automatically limit choice. The more order the less choice.

72

The Chromosome Does Not Need Selection to Conserve, Innovate and Explore

The chromosome contains within its boundaries all the mechanisms necessary to achieve these three processes, or put another way, the chromosome has such a molecular make-up that it cannot do anything but preserve, innovate and explore.

How does the chromosome preserve its sequences? How does it create new ones? How does it change its genetic pathways into new functional alleys? In other words, what are the mechanisms underlying these three basic processes. They turn out to be regulated by the internal organization of the chromosome in collaboration with the molecular pathways of the cell.

73

Repair Mechanisms Ensure the Maintenance of Order by Occurring at Different Molecular Levels — The Production of DNA, RNA and Protein Are Under Different Types of Control

Repair mechanisms are at the basis of chromosome and cell order. T.A. Brown (1999) puts it in clear language: "In view of the thousands of damage events that genomes suffer every day, coupled with the errors that occur when the genome replicates, it is essential that cells possess efficient repair systems. Without these repair systems a genome would not be able to maintain its essential cellular functions for more than a few hours before key genes became inactivated by DNA damage. Similarly, cell lineages would accumulate replication errors at such a rate that their genomes would become dysfunctional after a few cell divisions."

The chromosome has created, not one, but a series of mechanisms, by which it has ensured the maintenance of its original DNA pattern. However, it did not confine its intervention to this level. The DNA message, in the form of RNA, could easily distort the primary script. The final protein, obtained from this RNA, could also misrepresent the RNA message. The result is that the chromosome extended the repair and adjusting capacity to the RNA and protein levels, imposing strict order at every functional step.

74

Without DNA Repair No Human Would Exist

A human is the result of a single fertilized egg that becomes transformed into an adult organism consisting of 10^{13} to 10^{14} cells (Hood, 2002). To produce this cell mass an equally impressive number of cell divisions and of DNA replications has to occur.

If these processes were left at the mercy of selection no human being would exist.

As Alberts *et al.* (1994) have pointed out "If left uncorrected, spontaneous DNA damage would rapidly change DNA sequences". Thermal collisions with other molecules would lead to significant gene changes. It is known that about 5,000 purine bases are lost per day from the DNA of each human cell because of thermal disruption of certain atoms. This and other accidental base changes in the DNA molecule are eliminated by DNA repair.

There are different categories of DNA repair: (1) Direct repair converts each damaged nucleotide to its original structure. (2) Base excision is obtained by its removal and resynthesis of the DNA at this site. A similar mechanism does not remove the damaged base but acts on larger areas of DNA. (3) Mismatch repair corrects errors of replication by excising a piece of single-stranded DNA, containing the distorted sequence, and filling in the resulting gap. (4) Nonhomologous end-joining leads to the repair of double-strand breaks (Brown, 2007) (Fig. 4.3).

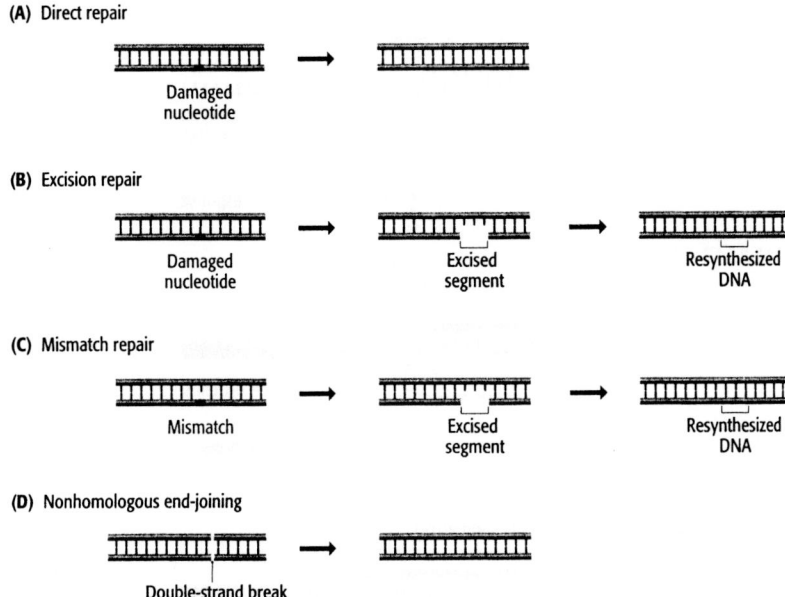

(A) Direct repair

Damaged
nucleotide

(B) Excision repair

Damaged
nucleotide

Excised
segment

Resynthesized
DNA

(C) Mismatch repair

Mismatch

Excised
segment

Resynthesized
DNA

(D) Nonhomologous end-joining

Double-strand break

Fig. 4.3   Four categories of the DNA repair system which allow the chromosome to correct itself.

(A) Direct repair. (B) Excision repair. (C) Mismatch repair. (D) Nonhomologous end-joining. In all cases the damaged DNA is converted to the original structure or modified into a viable one.

Other enzyme pathways may be involved in the restoration of the original function. DNA polymerases also participate in the proofreading mechanism that ensures that DNA replication occurs with a minimum of mistakes. Proofreading is the activity possessed by some of these enzymes which enables them to replace a misincorporated nucleotide during DNA copying. The fidelity of copying is of the order of one error in every 10^9 base pair replications. A protein, called p53, that exerts a tumor-suppressing function, has been found to facilitate DNA repair (Tanaka *et al.*, 2000; Lozano and Elledge, 2000). The importance of DNA repair in medicine became evident when several inherited human diseases were linked to defects in the repair process.

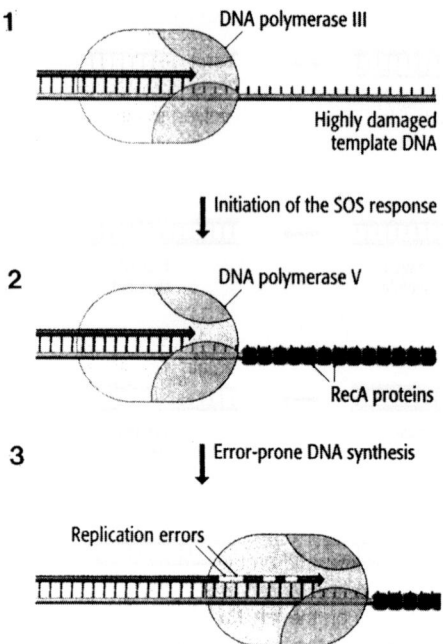

Fig. 4.4 Another way the chromosome uses to ensure its survival. The so called "SOS response" in the bacterium *E. coli.*

DNA can be damaged by mutagens or ultra-violet radiation, becoming partly single stranded (1). The SOS response results in the RecA protein coating the damaged strand (2) allowing DNA polymerases III and V to add the missing strand (3). The replicating errors are not eliminated by this process, but the chromosome is able to survive under adverse conditions, since its DNA is now double stranded. Later, other repair mechanisms intervene to correct the errors that have remained in the DNA.

Another way the chromosome uses to ensure its survival is the "SOS response" in the bacterium *E. coli.* DNA can be damaged by external agents such as chemical mutagens and ultra-violet rays, becoming partly single stranded. In the SOS response the RecA protein coats the damaged strand, allowing DNA polymerases III and V to add the missing strand. The replication errors are not eliminated by this process, but the chromosome is able to survive since it is now double stranded. Later, other repair mechanisms intervene to correct the errors left in the DNA (Fig. 4.4).

RNA Integrity Which is an Obligatory Condition for Normal Cell Function is Maintained by Another Type of Repair

The flow of genetic information in almost all living organisms, starts with DNA that makes RNA which in turn produces protein. Damage to any of these molecules can impair the normal information flow. The cell organization is so focused on ensuring order, that repair also occurs at the next level — that of the RNA.

The mental fixation on the importance of DNA, has been so widespread among biologists, that repair mechanisms at other levels were considered unlikely to exist. As pointed out by Begley and Samson (2003) it has long been known that cells repair chemically or physically damaged DNA, but "the discovery that damaged RNA can also be repaired may come as a surprise. What's more, some of the same enzymes are involved."

DNAs, RNAs and proteins are known to be chemically damaged by the aberrant addition of methyl groups (CH_3) to their molecules.

In *E. coli* a new mechanism has been discovered by which the bacterium repairs its methylated DNA. This is achieved by the demethylation of damaged bases which is catalyzed by an enzyme called *AlkB*. Significant is that the DNA of humans is also repaired by proteins related to this enzyme. The same mechanism being present at the two extremes of organism evolution. Of interest is that this repair extends to RNA. Moreover, the same enzyme (*AlkB*) which is involved in the repair of DNA participates in the repair of RNA using the same demethylation reaction (Aas *et al.*, 2003).

RNA is essential for most basic cellular processes. It participates in key events, such as protein synthesis, which demands the collaboration of messenger RNAs, ribosomal RNAs and transfer RNAs. Moreover, a battery of small RNAs, which are too small to code for proteins, regulate a variety of other chromosomal processes.

The integrity of RNA must be maintained at all costs, otherwise there would be no cell that could function properly.

76

RNA Surveillance — An Additional Mechanism that Improves Safety by Creating Quality Control

RNA surveillance is actually a form of quality control of the RNA. Once this molecule has been made it moves across the nuclear sap to the pores of the nuclear envelope arriving finally in the cytoplasm. This is a long journey filled with difficulties and hazardous situations like any other uncertain adventure.

The scrutinizing system consists of molecular assemblies that specifically control the quality of the messenger RNAs and it involves numerous steps, both inside and outside the nucleus (Culbertson, 1999).

There is a first quality control at the level of the DNA which is performed by 10 proteins that mediate its decay. In this process the messenger RNAs are monitored for errors that arise during gene expression. Base substitutions often cause chain termination. This results in the building of nonsense RNAs. These RNAs, which code for protein fragments, are recognized and eliminated (Lykke-Andersen *et al.*, 2001).

The next step is the export of the messenger RNA. This is mediated by a transport receptor that binds to the messenger by means of adaptor proteins. This RNA-protein complex moves to the pores of the nuclear envelope. Here it interacts with a second type of

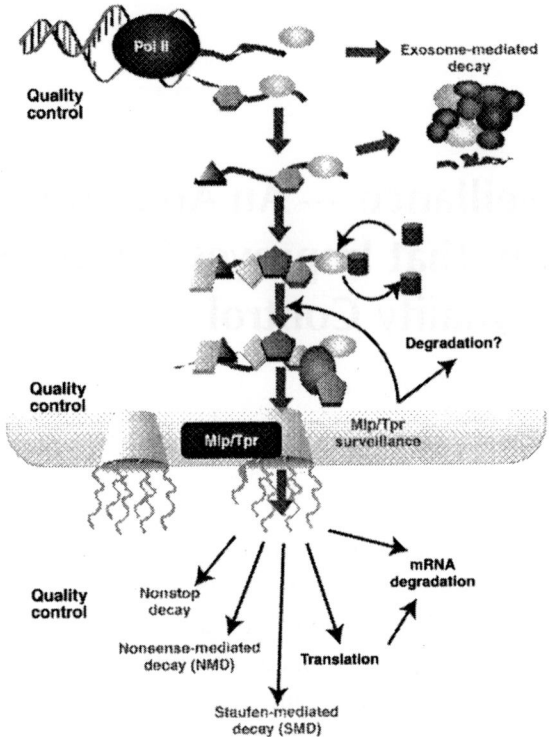

Fig. 4.5   The export of RNAs is quality controlled.

The quality control of the coding messenger RNA occurs at three separate levels: (1) Immediately after transcription on the chromosome there is an "exosome"-mediated decay directed by several proteins (top right). (2) Later the RNA gets associated with another protein which directs it to the pores of the nuclear envelope, where a second quality control occurs (middle bar). (3) On arrival at the cytoplasm (lower part of figure) the RNA goes through a series of decay processes that dispose of aberrant molecules. Damaged molecules are excluded from translation into proteins. It turns out that all nuclear RNA export is protein mediated.

proteins which are at the gates of the membrane. The messenger RNA assembly is then subjected to quality control by a nuclear surveillance mechanism so that aberrantly assembled RNAs are degraded before they are delivered to the cytoplasm (Cole, 2001; Stutz and Izaurralde, 2003) (Figs. 4.5 and 4.6).

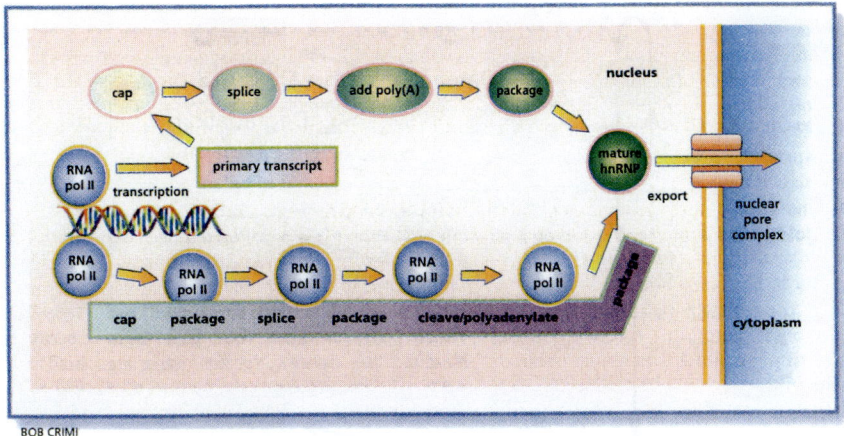

BOB CRIMI

Fig. 4.6   Messenger RNAs of higher organisms must be accurately processed before they can be exported from the nucleus to the cytoplasm.

Following transcription of the RNA from the DNA, the processing includes the following steps: 1) Capping of the RNA. 2) Splicing to remove introns. 3) Cleaving of the nascent transcript. 4) Addition of a poly(A)tail. 5) Packaging of the mRNA for export by forming a complex with several specific proteins and export factors. Once fully processed and packaged, the mRNA is exported to the cytoplasm for translation into protein.

RNA pol II = RNA polymerase II which catalyzes the transcription of messenger RNA from DNA (left part of figure). The upper and lower pathways illustrate two variants which lead to the same final result.

A third quality control takes place when the RNA arrives at the cytoplasm where several types of mechanisms dispose of it or degrade aberrant molecules (Fasken and Corbett, 2005).

But, not only the messenger RNA, but other types of RNAs, such as non-coding RNAs, have their nuclear export controlled by association with specific proteins during their journey to the cytoplasm (Cullen, 2003).

The final result is that no damaged molecules become available to build defective proteins. These molecular mechanisms, which impose functional coherence, have been conserved from yeast to humans (Vinciguerra and Stutz, 2004) (Fig. 4.7).

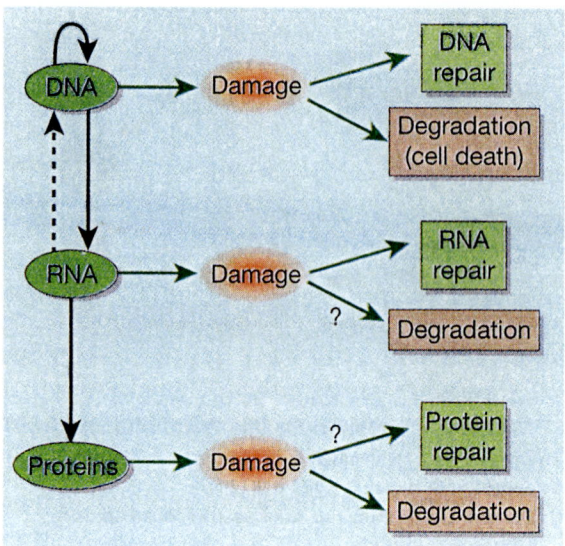

Fig. 4.7 How the major biological molecules counteract damage.

DNA is needed to make RNA, which in turn is needed to make protein. This flow of genetic information is found in almost all living organisms. Information can also flow back from RNA to DNA in the case of RNA viruses; and DNA molecules can be replicated leading to life's propagation. DNA is the ultimate repository of information and thus it is vital that it is repaired when damaged. It has been thought that RNAs and proteins are dispensable in part, because new copies can be easily generated. But RNA can also repair itself adding to the maintenance of order at another level. Protein repair is not yet established, but instead they use "molecular chaperones" that correct their assembly.

77

"Molecular Chaperones" Are Proteins that Ensure that a Correct Molecular Assembly Will Predominate

The name given by molecular biologists to these proteins is most appropriate because the Collins English Language Dictionary (1987) tells us that: "A chaperone is an older or married woman who used to accompany a young unmarried woman on social occasions, especially when there were men present". She made sure that men's advances were kept at a certain distance but at the same time helped to promote marriages of convenience.

The description of such a situation, most common in the society of the early 1900s, is dramatic for it tells us how women were under the control of other women for such a long time.

The main function of chaperones is to facilitate the correct assembly of proteins. To be noted is that they are not components of the assembled structures, i.e. they only help and supervise. Moreover, they do not convey information either for polypeptide folding or for the assembly of multiple polypeptides into a single protein. Instead, they function by binding to specific structural features that are exposed only in the early stages of assembly.

Like human chaperones, who inhibited economically disadvantageous marriages, the chemical chaperones inhibit unproductive assembly pathways that would lead to incorrect structures. Also like the respectable ladies, the molecular counterparts participate, but are not part of the self-assembly of proteins (Becker *et al.*, 2003).

78

How to Confuse Evolutionists — The Correction Can Function Backwards, Ancestral RNAs Can Restore the Original DNA Sequence

It is well known that DNA produces RNA and that this molecule carries its message to be later translated into protein. Following its formation RNA is released into the nucleus and later into the cytoplasm where it carries its own activity. It turns out that RNA has not only the capacity to repair itself, and to survey its own structure, but also to act in reverse order, being able to correct DNA, i.e. to rectify the sequence of its parent molecule.

A previously unknown way of reversing the chromosome sequences, has been revealed by the analysis of plants carrying mutations in a specific gene. In *Arabidopsis*, Lolle *et al.* (2005) have found mutant strains of this gene that yielded, instead of the mutated form, a normal progeny in a high frequency. This unusual situation was found to be due to a precise reversion, that restored the original DNA sequence, that had earlier mutated (Fig. 4.8).

The restoration of the normal condition was attributed to the presence of a template-directed process that made use of an ancestral RNA-sequence that had remained hidden in the cell.

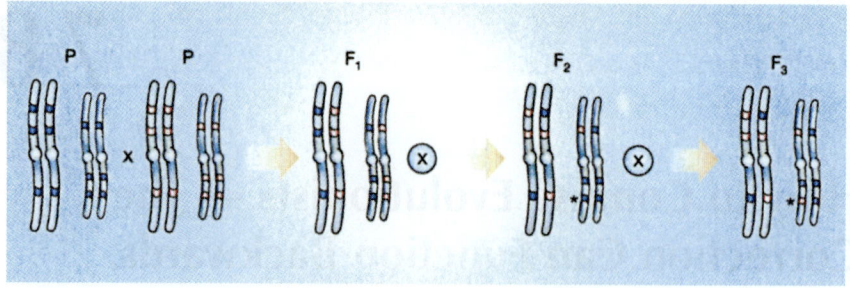

Fig. 4.8   A previously unknown way of reversing changes in DNA.

The analysis of *Arabidopsis* plants with a specific mutation led to the discovery of a reversion in their mutation process that cannot be explained by random mutation. Instead the progeny of the mutant plants can recover DNA variants in later generations that have come from one of their great-grandparents, even if their immediate parent did not contain the variant. Two parents (P) with different DNA sequence variants at several positions (colour-coded) in their chromosomes are crossed. In subsequent generations (F1 and F2) plants are self-fertilized (indicated by a circled cross). In the F3 generation there is reversion (asterisk) to a sequence originally present in the grandparents, but not in the immediate parents. This has been observed at several chromosome sites at surprisingly high frequency in this *Arabidopsis* mutant. The restoration of the normal condition has been attributed to the presence of an ancestral RNA.

In a comment on their work Weigel and Jürgens (2005) concluded that "There is no doubt that the reversion rates observed by Lolle and colleagues in the plant *Arabidopsis* cannot be explained by random mutations". No better way could be found to confuse evolutionists. The chromosome reverses the rules using RNA to change DNA.

79

Innovation by Creation of New
Gene Sequences

The discovery of the split gene carried with it many other unexpected findings. The immunoglobulin gene is created by the chromosome, by simply using its trivial molecular tools. Two DNA sequences, which in the germline of the mouse had no function, (i.e. did not transcribe RNA) and as such could not be recognized as genes, are put together, by means of a rearrangement, in the somatic tissues to become active and build the immunoglobulin gene. The two segments do not come completely together but are separated by an intron consisting of a specific number of bases. The gene is created by using such a simple tool as a rearrangement but that does not inhibit the chromosome from doing it with the highest precision since an intron of a specific length is formed each time the gene is built (Tonegawa *et al.*, 1978). The same phenomenon occurs in humans, three chromosomes being involved in the rearrangement that results in the formation of the active gene. The recombination events are directed by a *recombinase* enzyme complex (Strachan and Read, 2004).

Redundancy and amplification are also innovative processes. The chromosome is able suddenly to produce hundreds of copies of the same gene sequence. It may add these copies to itself longitudinally (redundancy) or laterally (amplification). The result is not only quantitative, as one would think i.e. higher and more rapid production of a given RNA or other molecule, but qualitative.

Once there are hundreds or thousands of copies of a gene at a given site, instead of the normal number of only a few copies, the neighbouring genes in the same chromosome, or genes in other chromosomes, may be affected in their functions. Nucleolus organizer regions, which contain many copies of ribosomal RNA genes, are known to interact with each other whether they are in the same or in different chromosomes (Pikaard, 2000; Ferrari *et al.*, 2006).

80

Exploration is Achieved by Change of Genetic Pathways into New Functional Alleys

The chromosome actually explores new solutions all the time. Chromosomes have a remarkable property that one tends to take for granted because it occurs so often, yet it has a high genetic significance. DNA can be cleaved and reunited and this allows chromosomes to rearrange. Many types of rearrangements are known to occur, classified as inversions, duplications, translocations and deletions, and they may involve thousands of nucleotides or only a few. The rearrangements in themselves do not need to be innovative, but they have consequences which are. In many well-established cases they produce changes in function, and in the type and rate of mutation, of neighboring or far-situated DNA sequences.

The new functional alleys are established by mechanisms which have been designated before as position effects. Correct tissue specific gene expression can depend on sequences located hundreds of thousands of DNA bases away from the coding gene. Several human genes, involved in disease, are examples of such long distance position effects (Kleinjan and van Heyningen, 1998; Bell *et al.*, 2001).

By means of rearrangements the chromosome creates new constellations of genes which did not previously exist and new functional solutions. The important point is that it does it within its own boundaries and with the help of its own molecular tools. It needs nothing else than immersion in a cell from which it acquires the necessary molecules.

81

How Plasmids and Accessory Chromosomes Evade Selection

Plasmids are minor chromosomes which consist of double-stranded DNA and occur mainly in bacterial cells. They are independent units because they have their own system of replication and contain their own genes which are not usually encoded by the main bacterial chromosome.

Plasmids evade selection by two main efficient methods. The first is when the plasmid is transferred from one bacterial cell to another during conjugation. The plasmid can carry genes that enable the recipient cell to survive at the expense of the donor cell. As Novick (1980) points out "the plasmids have evolved the ability to survive regardless of the fate of their host species — something that would be inconceivable, within the framework of evolution through natural selection, for an element that was merely a component of a particular organism's genome". The cell dies but the minor chromosome, which should be completely dependent for its existence on the cell, manages to survive independently of the fate of the cell.

Still more significant is the second property displayed by plasmids. Their genes are arranged in such a way that they ensure both genetic stability and genetic plasticity. Plasmids can acquire new genes and rearrange the old ones. This allows them to maintain a store of genetic information consistent with the needs of their current host organism. This results in the cell being able to surmount many difficult physiological situations. The plasmids are chromosomes with great possibilities of genetic innovation, which further increases their ability to evade selection. This innovative ability is a direct product of their molecular construction.

Accessory chromosomes of higher organisms are usually smaller and variable in number. They exist at the side of the large chromosomes of the normal complement. They evade selection in a way comparable to that of the plasmids. By having their own mechanisms of increase and their own genetic effects they are kept in the genome population as a source of genetic innovation. Among other effects they are known to significantly change chromosome pairing and recombination (Sheidai *et al.*, 2006).

The accessory chromosomes allow the eukaryotic cell to cope with new environments without changing its main collection of chromosomes, just as the bacterial cell uses its plasmids in a similar way keeping its main chromosome relatively unaltered.

82

There Are Genes Which Are Able to "Cheat" Natural Selection

Crow (1979) presented a clear cut example of genes which, as he stated in his work, "cheat selection". This is a gene located near the centromere of chromosome II of the fly *Drosophila melanogaster* called the segregation distorter. This gene, working in collaboration with two other genes and an enhancer, changes a chromosome segregation from the 50: 50 normally obtained to a 99: 1 ratio. This event modifies radically the chromosome distribution within the cell. The collaboration between these "distorter genes" alters in a dramatic way the behavior of the chromosomes.

The chromosomes themselves can change at will their own assortment from fully independent to non-independent. This is done by molecular processes which are inherent in the organization of the chromosome.

83

Sensing Mechanisms Are Used by the Chromosome to Adjust Gene Number and Switch on Genes that Improve Survival

"Sensing mechanisms in chromosomes" seems to sound like pure mysticism but it has become a standard expression in molecular biology. The chromosome has no need of higher powers. It contains within its atoms the properties that allow it to sense its organization.

Ritossa (1973) and Tartof (1974) found in the fly *Drosophila* that gene magnification was dependent on a sensing mechanism capable of recognizing the number of ribosomal RNA genes present in the chromosome and of introducing the necessary adjustment obliging the gene level to return to the original one. A wild-type locus consists of 130–300 ribosomal genes. If the genome contains less than 130 genes for ribosomal RNA, its synthesis is inadequate and the mutated condition develops. A reversal to the normal number of genes occurs in the progeny of these abnormal flies by a rapid accumulation of ribosomal RNA genes bound to the chromosome. The increase is inheritable and causes in one, or a few generations, the achievement of the wild-type number. Tartof thinks that the chromosome is capable of "sensing" a deficiency in the number of ribosomal RNA genes and of introducing the necessary adjustment.

The agent of the sleeping disease in humans is the protozoan *Trypanosoma*. It is a parasite which evades the immune system of the host and can switch on new genes that code for its surface antigens. In this way it rapidly changes the properties of its surface coat allowing it to survive. It simply puts on a new dress confusing the immune system. These variable genes are expressed when two events occur: (1) an extra copy of the gene is produced and (2) it is translocated to a telomere region. Only the copy that is located near the telomere produces the messenger RNA of the new coat glycoprotein. The gene may have different positions along the chromosome but the telomere sequences are necessary for expression (Donelson and Turner, 1985). Evans and Lundblad (2000) studying the regulation of telomerase access to the telomere concluded that: "DNA binding proteins are part of a length-sensing mechanism that can discriminate the number of duplex-binding proteins bound to the telomere".

The sensing processes allow the chromosome to maintain its integrity and to avoid extinction.

84

The Multitude of Protective Mechanisms Devised by the Chromosome "Prohibit Natural Selection"

Cairns (1998) who studied extensively mutation in bacteria points out that it is customary to think of mutations as being driven by chance because these events are attributable to the natural instability of nucleic acids and the inherent imprecision of their copying enzymes.

He notes that every living organism is protected against changes in DNA sequence, and that there are four main such mechanisms at work: (1) Instead of nucleic acids being single-stranded they are usually double-stranded, a feature which minimizes error. (2) There are the well known proofreading polymerases that correct base mistakes. (3) The equally well known DNA repair mechanism ensures correct base inclusion along the DNA strands. (4) There is moreover a molecular monitoring of the developing embryo which only allows the formation of a coherent organism which is essentially alike its parents.

Cairns makes this last point more precise when he adds: "The invention of multicellularity has been accompanied by the development of mechanisms that prohibit natural selection". During evolution single cell organisms became transformed into huge aggregates of millions of cells. These were produced starting from a single fertilized egg. Without the protective mechanisms that ensure molecular interactions between cells, at every step of chromosome and cell copying, nothing comparable to a living organism would have emerged.

Let us now describe how selection is foreign to the embryonic development of a multicellular organism and to the emergence of its organization.

The Aggregation and Cell Adhesion of *Dictyostelium* Cells Follow the Same Chemical Solutions Employed by Embryos of Higher Organisms

The slime molds, such as *Dictyostelium*, pass through a unicellular stage of amoebas that feed on bacteria. Subsequenly, these amoebas aggregate to form a fruiting structure that produces spores. The organism displays an incredible feat of organization. Its intriguing development has been studied in the utmost detail at the genetic and chemical levels (Kuczmarski and Spudich, 1980; de la Roche *et al.*, 2002) (Fig. 6.7).

Thousands of single cell amoebae aggregate into a single multicellular body. Using microscope cinematography the individual cells can be seen moving at the speed of 20 microns per minute. The cells are guided by a cyclic adenosine 3′, 5′-monophosphate (cAMP) that is recognized by cell receptors leading to specific gene transcription and motility.

Following aggregation the cells adhere to each other. This adhesion is mediated by a glycoprotein (a combination of a sugar with a protein) following the activation of a messenger RNA. This is in turn followed by a second and a third series of adhesion molecules which transform the individual cells into a coherent organism. Similar chemical events are involved in the formation of embryos in complex organisms (Ginger *et al.*, 1998; Loomis and Insall, 1999).

The Egg is a Storehouse of Information, Prepared by the Mother's Chromosomes — This Guarantees the Formation of an Identical Body Pattern

According to Gilbert (2000) the treasure chest of the egg consists mainly of molecular information that will charter its future development.

(1) There are proteins that will nourish the developing embryo before it can get them in large amounts. These molecules have travelled a long way. Some were produced in the liver of the mother and reached the egg through her blood.

(2) Following fertilization, protein synthesis starts on a large scale, but this is achieved by ribosomes and transfer RNAs that existed already in the egg.

(3) The sea urchin treasure chest contains 25,000 to 50,000 different types of messenger RNAs, which carry the instructions to build the embryo. The time at which they will start their performance is fixed. They are kept dormant until they are activated by the correct ionic signals. The functions encoded by these RNAs are many and highly diversified. They are the keys to the initiation and maintenance of crucial events such

as: cell division regulation, cell movements, formation of the spindle necessary in division, chromatin increase, cell adhesion, DNA and protein synthesis, determination of cell differentiation and others. There is practically no basic event, involved in the determination of the embryonic pattern, that is not carried out by these RNAs.

(4) Eggs are usually exposed to an adverse environment. It turns out that they contain ultra-violet filters, as well as DNA repair enzymes, that protect them from the sun's rays. The yolk of bird eggs may contain antibodies that are a defense against infection.

(5) The initial stages of the development in *Drosophila*, and sea urchins, do not require the presence of a nucleus in the cell, demonstrating that the oocyte's molecular blueprint is the sole director of the initial order needing no information from the chromosomes of its nucleus (Edgar *et al.*, 1994).

The predictive capacity of the chromosomes of the mother could not be more evident. They produced the thousands of molecules which created the treasure trove of information which ensured the future of the species by only allowing the formation of an ordered body pattern identical to that of the previous organism.

87

The Genetic Code Does Not Contain Direct Information to Produce a Coherent Organism — This Lies in the Hands of Other Molecular Processes that Charter Development by Building a Road Map

The genetic code located in the chromosome's DNA is responsible for the information that results in the organism's different proteins. But this information could not alone lead to the establishment of any coherent organism, because it does not contain a "road map" showing: 1) The pathways to follow, as the cell number increases. 2) The exact sites where new structures are to be built. 3) The time at which they are to be started and finished. 4) The spatial relations between the components of the body pattern.

The plethora of messenger RNAs produced by the mother's chromosomes function as a pattern code that decides the main architecture of the new organism which originated from a single egg cell. But, the messenger RNAs are as many as several thousands. Such an enormous number would only lead to utter confusion. A key phenomenon is that they are dormant, i.e. remain inactive awaiting a signal. The source of such signals were before unknown but now it turns out that they are given by small RNAs.

88

Minute Cell RNAs, that Previously were Despised, Turn Out to Coordinate Messenger RNAs

The finest perfumes are found in the smallest bottles. Big is not necessarily best, but in science it took time to discover this.

The cell's small RNAs were despised. They were too small to produce a protein and as a consequence were considered of no interest. It turns out now that they are among the keys to the order of gene function and that they decide the fate of the messenger RNAs that are stored in the cell. The understanding of gene regulation in animals, plants and fungi, has changed profoundly due to the action of a vast new world of tiny regulatory RNAs. They have been called "short interfering RNAs" and "micro RNAs". Their regulatory role is known as "RNA interference".

The micro RNAs may have a length of 21 to 22 nucleotides and arise from non-protein coding genes (Carrington and Ambros, 2003; Denli and Hannon, 2003). They regulate gene expression by binding to complementary messenger RNAs. This is possible because the small RNAs have the same base sequences as their target messengers. This sequence recognition triggers: 1) messenger RNA elimination and 2) the arrest of messenger RNA translation into protein. These events are critical in shaping the road map of the emerging embryo (Matzke and Matzke, 2003) (Figs. 3.10, 4.5 and 4.6).

Hence, the genetic code, which is responsible for the production of messenger RNAs and micro RNAs, participates only indirectly in the creation of the road map, or pattern code, that decides the architecture of the organism.

89

The Mechanisms Responsible for Coherence and Order Have Been Experimentally Demonstrated

It is the cell surfaces, with their protein "anthenae", that are mainly responsible for the recognition between cells and the establishment of their affinities. Each cell type has a different set of proteins on its surface.

Experiments were devised, in which amphibian tissues were dissociated into single cells and the resulting cell suspensions were combined in different ways. By using embryonic cells from different species, which had different colours and sizes, their fate and distribution could be followed in the new aggregate (Townes and Holtfreter, 1955).

The results were striking: 1) In the ensuing mixture the cells became spatially segregated and each cell type moved to its own particular region. 2) The final positions of the cells in the novel aggregate became those that they had in the original embryo. 3) In no case did the cells become randomly distributed. They were able to sort themselves out into their original embryonic locations.

Similar reconstructions of aggregates, from later embryos of birds and mammals, led to the same result. The cells were also able to reconstruct the organization of the original tissue (Moscona, 1952).

The ordered movement was followed by an equally ordered adhesion between the cells of the embryo. This was mediated mainly by the proteins cadherins and catenins, different cell types having different adhesion molecules. Following the removal of the protein E-cadherin, the cells could not hold together. This protein was the molecule responsible for the binding process (Godt and Tepass, 1998).

The billions of cells that make up the human brain, find their correct position, in relation to each other, by the same molecular mechanism. Cadherins have been implicated in providing an adhesion code for neural circuit formation during wiring of the brain (Gumbiner, 2005).

90

The Drastic Reshapings that Occur in the Embryo Are Directed by Specific Proteins

After the egg has been transformed into an aggregate of thousands of cells a new stage ensues, called gastrulation, in which all these cells are rapidly reorganized into the shape of the new organism.

Embryologists describe this novel stage as the product of the following events:

1) The first step is characterized by intimately coordinated cell movements.

2) These involve the entire embryo, the cells migrating from the surface to the interior as well as in other directions.

3) The migrating cells have an address that obliges them to stop their movement at the precise site.

4) Three embryonic axes of symmetry are formed in the embryo which are the determinants of the new body order. In *Drosophila* the orientation of these axes is determined by the position of the egg within the other cells of the ovary. The formation of the axes is dependent on the presence of four protein and several RNA gradients which occur along specific directions of the embryo.

5) What a cell becomes later in animal development depends upon its position in the embryo. Its fate is determined by interactions with neighboring cells (Fig. 4.9).

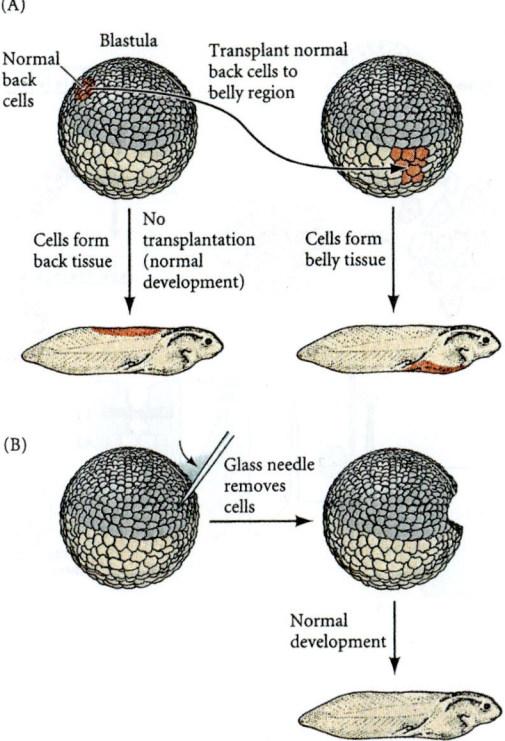

Fig. 4.9 What a cell becomes later in animal development depends upon its position in the embryo.

(A) A cell's fate is determined by interactions with neighboring cells. When cells of the back part of an embryo, at the blastula stage, are allowed to be in their natural position, they continue to be present at the back in the amphibian tadpole. However, when the back cells are transplanted to the belly region of the blastula they get a belly location in the tadpole. (B) If cells are removed from the embryo with the help of a needle, the remaining cells can regulate and compensate for the missing part.

In the first case, the cells from back tissue were transformed into belly tissue as a result of transplantation. In the second case, the removal of a group of cells did not alter the formation of a normal individual in amphibians. In both situations the original pattern could not be altered.

6) Drastic morphological changes occur in the embryo which give it its novel shape. The mechanisms that guide these reshapings are now understood. The proteins called cadherins have been found to have a function, not limited to cell adhesion, but

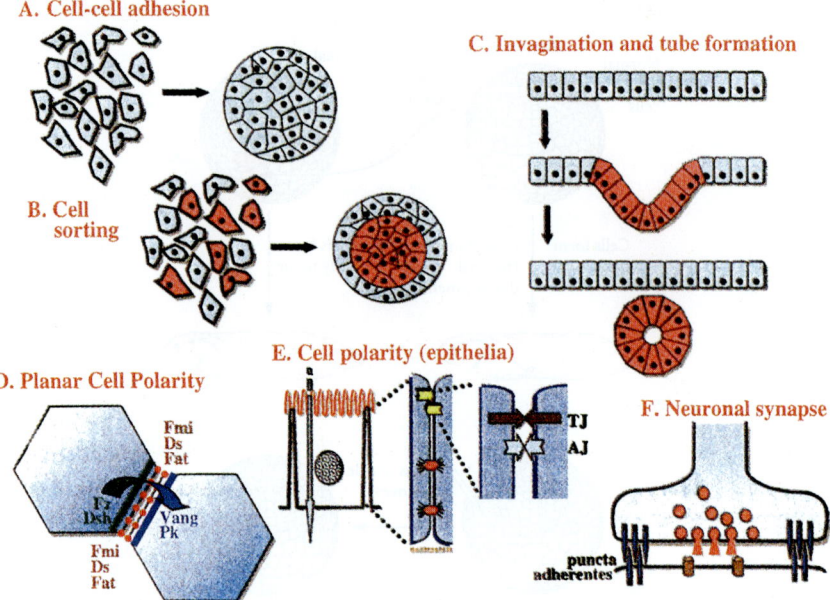

Fig. 4.10   The diversified roles of cadherins during embryonic development.

The dramatic transformations that change an embryo into a newborn were most difficult to explain, because they were so different and so innovative. Suddenly they turn out to be directed by a family of proteins. The morphogenetic capacity of the cadherins includes: (A) Cell-cell adhesion. (B) Sorting out of mixed cell populations. (C) Coordination of cell movements leading to invagination and tube formation. (D) Planar cell polarity. (E) Activation of cell polarity in epithelia, such as skin. (F) Last but not least, they are implicated in neuronal synapse, a critical event in locating the millions of brain cells in their correct positions. These are crucial transformation steps in the embryo for which before there was no known molecular mechanism.

which extends to multiple aspects of tissue morphogenesis. These include: cell recognition and sorting out, boundary formation, coordinated cell movements, induction and maintenance of cell polarity. Such diversified activities are carried out by over 100 members of the cadherin family (Halbleib and Nelson, 2006) (Fig. 4.10).

This biological scenario is so unique that it led Gilbert (2000) to point out: "Somehow, the embryo knows that some organs go

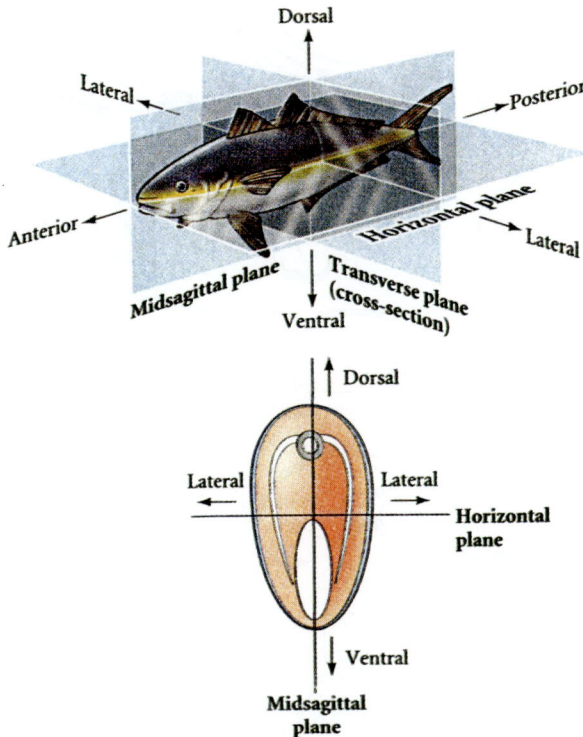

Fig. 4.11 The axes that become evident in the adult animal are present in the embryo. These are the foundations of the future body plan.

Representation of the axes of a bilaterally symmetric animal such as a fish. A single plane, the midsagittal plane, divides the animal into left and right halves. Cross sections are taken along the anterior-posterior axis. Humans are subjected to the same type of symmetry consisting of the 3 axes: anteroposterior, dorsoventral and right-left (lateral). "Somehow, the embryo knows that some organs go on one side and other organs go on the other" (Gilbert 2000).

on one side and other organs go on the other". The result of this axis specification is the determination of organ location along the body. Some organs are formed on both sides of the central axis, such as the lungs and the kidneys, but others are only organized on the left side, as the heart and the spleen, or only on the right side like the liver. In sea urchins some of these axes are specified as early as fertilization and are due to maternal influence (Henry and Raff, 1990) (Figs. 4.11 and 4.12).

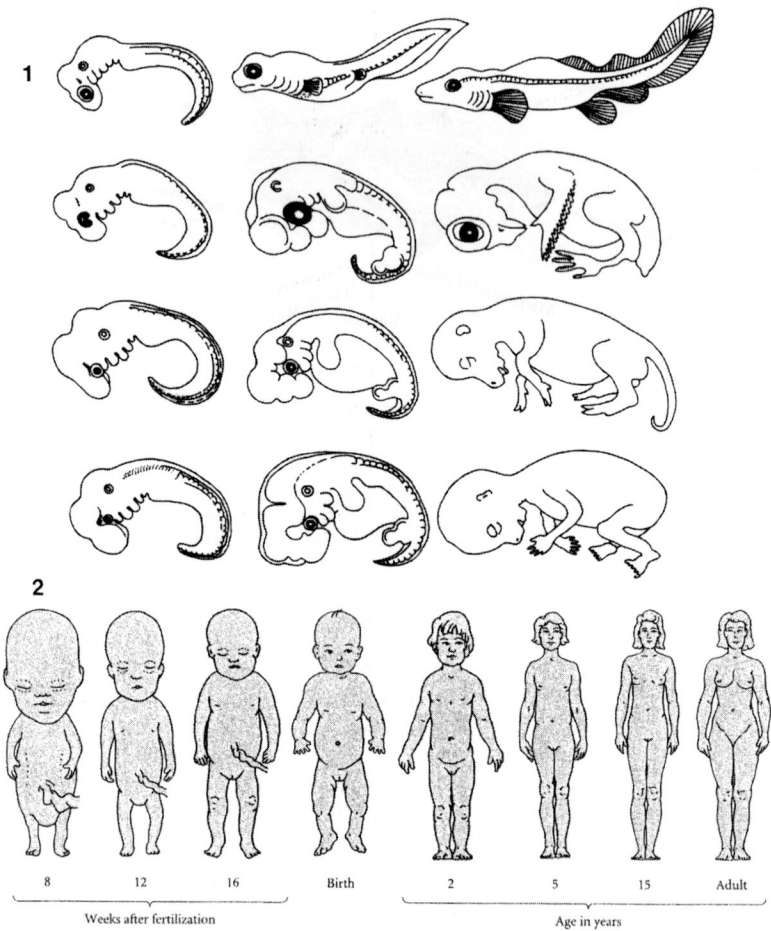

Fig. 4.12 Shapes and proportions of animal bodies change in predictable ways.

1. Comparison of the embryonic development of a bony fish, a bird (chick), a calf and a human. Progressive development from embryos (left column) which are, from the beginning, essentially similar.

2. Not all parts of the human body grow at the same rate. The embryo's head is exceedingly large in proportion to the rest of the body. After the embryonic period the head grows more slowly than the torso, hands and feet, and the arms and legs grow at a faster rate than the torso and head.

91

Cells of One Group Change the Shape, Mitotic Rate and Pattern of Their Neighbors

A complex organ, such as an eye is constituted by groups of cells with different functions but surprisingly these lead to a coherent final result denoted vision. Light, is transmitted by the transparent cornea, is focused by the lens, a muscle intervenes in changing its diameter, it reaches the retina, which sends the information to the brain, where it is recognized as form and colour. The five stations which participate in this mental process are constituted by totally different tissues, yet their collaboration is most harmonious otherwise there would be no vision.

How is this harmony and coherence achieved? The answer is to be found in simple chemical interactions in which separate groups of cells, change the behavior of adjacent ones, by modifying their shape, rate of division and pattern. This has been shown by experiments in sea urchin embryos. Of the three germ layers of an embryo the outer one is called ectoderm and the middle one mesoderm. They give rise to different tissues and organs. Isolated pairs of cells from the cap region give rise to both ectoderm and mesoderm. However, when aggregated with other cells from the cap, they are inhibited to form mesoderm. Their presence restricts the potencies of their neighbors (Henry *et al.*, 1989).

In sea urchins the number of skeletal mesoderm cells is fixed but can be regulated. If 60 cells are removed from this region of the embryo, an equal number is produced which replaces the loss. If only 20 cells are taken away about 20 are replaced. The original pattern is restored by reestablishing the correct number (Khaner and Wilt, 1991).

92

The Chromosome Has Made Sure that the Organism Not Only Protects Itself from Inner Errors But Also from Outer Enemies

The chromosome and the cell have devised all those protective molecular systems that allow them to follow their own path relatively undisturbed. However, such a feat would not be, in itself, sufficient to continue to inhabit the planet. The inner enemies — the undesirable deviations — most of them had been circumvented. But there were the external enemies, that had been waiting outside for an attack. These had also to retract because the chromosome and the cell answered their intrusion by developing the immune system that counteracted their pernicious action.

The earliest form of defense against infection from foreign bodies, emerged with the antimicrobial peptides. These were produced by chemical receptors that recognized pathogens ensuring host defense. Such receptors have been conserved over a long evolutionary period since they are found in flies, mice and humans where they recognize pathogens derived from bacteria, yeast, viral RNA and bacterial DNA (Ganz, 2003).

Another innate form of immunity in animals consists of the phagocytic cells that scavenge pathogens which manage to enter the blood stream. The incorporation of unicellular amoebae, which have also the ability to ingest foreign particles, is considered a possible origin of this defense system (Smith, 2001).

Proteins containing immunoglobulin-like domains are ubiquitous throughout the plant, animal and bacterial kingdoms. They have many different functions, one of them being an essential part of the immune system. Significant is that these molecules have no intermediate molecular forms, their peculiar features have arisen from the very beginning indicating that they were ready-made (Hoffmann *et al.*, 1999; Agrawal, 2000).

In higher organisms, such as the vertebrates, the immunoglobulin genes play a central role in organism defense. As usual, selection was invoked to explain the enormous range of cellular diversity present in the immune system, but that was before this process was studied at the molecular level. It now turns out that, on the contrary, the cell interactions are orchestrated by exact molecular recognitions carried out by protein receptors. This process leads primarily to the defense of the individual against deadly injuries, but secondarily to two equally relevant phenomena: programmed cell death and cell memory (Janeway *et al.*, 2005).

Cell Death is as Programmed as Cell Life

We tend to think that an organism starts as an egg, soon becomes an embryo, and later is transformed into an adult, by a continuous increase in cell number due to innumerous divisions. This is only one side of the process. A parallel event taking place at the same time is cell death. Significant is that the two opposite events have features in common — both are strictly programmed and are of similar proportions. Construction and destruction are the two faces of the same coin.

The number of somatic cells that build the simple worm *C. elegans* are exactly 959 (Brenner, 1974). Of these 131 are programmed to die if a signal does not intervene to prevent this event. It is the *ced-9* gene that regulates the choice between life and death at the cell level (Hengartner *et al.*, 1992). In humans, the astronomical figure of 10^{11} cells die, in each adult, each day, and are replaced by other cells with similar functions. In the uterus and the intestine thousands of cells are being replaced every day (Adams and Cory, 1998). Cell replacement is due to the production of chemicals which enhance the division of neighbor cells in specific directions.

Both in life and in death the instructions sent out by groups of cells are active even in unrelated species. Chick tissue responds to chemical induction by mammalian cells. Moreover, similar proteins are utilized in generating the eye (or the heart) of a fly and of a human. These proteins may traverse a filter placed between two tissues demonstrating their ability to diffuse and travel long distances. One of these inducer proteins is activin (Gurdon *et al.*, 1994).

Programmed cell death is also an obligatory event during flowering plant development. An example are the nucellar cells of the ovule which undergo death which is genetically controlled (Luigi *et al.*, 2006).

94

Cells Can Commit Suicide but Amoebae Are Potentially Immortal

The one cell amoeba one sees today, under the microscope, has no dead ancestors. The two resulting cells, of a division, are both ancestor and offspring since they are siblings. An amoeba only dies when it is eaten or suffers an accident.

On the other hand cells can commit suicide. In the simple green alga *Volvox*, a specific gene *regA* is known to regulate cell death. The suicide cells, are part of the general pattern, their nutritional contents are used to feed others, allowing an extremely rapid development (Kirk, 1999).

Cell suicide is also constructive in other unexpected ways. Our eyes are an exceptional organ due to controlled cell destruction. The lens of the eye is the only transparent tissue in the human body. This transparency is achieved by a self-destruction program of cell contents which eliminates the nucleus and other organelles, but stops just before killing the cell itself. This degradation leaves nothing but an outer cell membrane and a thick solution of special proteins, called crystallins, which fill the whole cell. The result is a structurally empty but viable cell that transmits light (Bassnett, 2002; Dahm, 2004). The same situation occurs in the red blood cells of the human body which are fully functional but lack nucleus, mitochondria and most internal membranes.

Both the Cell and the Chromosome Have an Unfailing Memory

Cell memory was a concept foreign to biology until the 1950s. No one would dare to consider the occurrence of the ability to retain, store and recall information at the cell level. The evidence obtained since then has established memory as one of the properties of the cell and the chromosome. They have the ability to learn and to use subsequently the acquired knowledge.

Cell memory was initially documented in the defense of organisms against infection. The immune system of humans has the ability to respond more rapidly and effectively to agents of disease that have been encountered previously. Children that are exposed to the measles virus through infection acquire long-term protection from measles. This immunological cell memory is found in other acute infectious diseases and persists after the exposure to the virus has ceased. By one month after immunization, cells that carry the memory of the infection, called "memory cells", are present at their maximal levels in the body. These levels are then maintained for the life-time of the individual (Janeway *et al.*, 2005). The fixation and perpetuation of novel properties is achieved by groups of genes being both activated and inactivated, the new gene situation being locked with the help of histone proteins (Turner, 2002) (Fig. 4.13).

As expected, it is the chromosome, with its chemical manipulations, that is the structure responsible for the cell's memory. But this capacity turns out to have still more intriguing aspects.

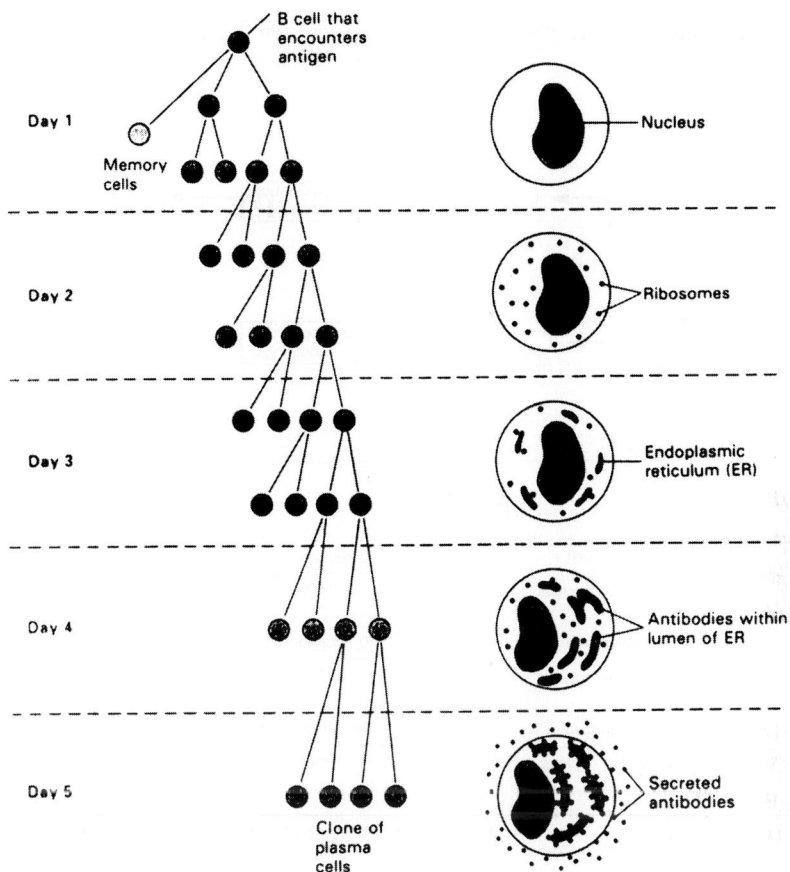

Fig. 4.13   Cell memory.

Memory cells develop in a line of so called B lymphocytes, that encounters antigen. This leads to the formation of a clone of plasma cells. Cells remember their state of gene activity through divisions, which allows the immune system to recognize earlier molecular events.

When normal human fibroblasts were grown in tissue-culture it turned out that aging of normal cell lineage was an innate property of the cells (Hayflick, 1980). The cells also exhibited a 'memory'. If the cells were frozen in liquid nitrogen at the 20th cell population doubling and later taken to room temperature they would divide 30 more times but after that they would die. If the experiment was repeated but this time at the 10th cell doubling the cells would

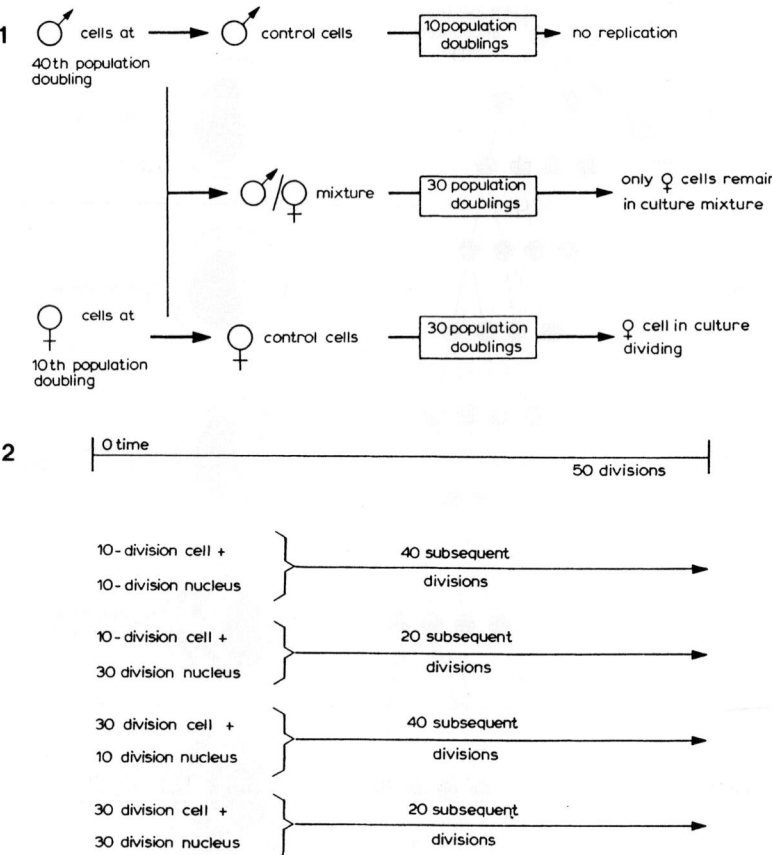

Fig. 4.14   The chromosome is able to count the number of times it has divided.

1. Limit of replication of normal human cells in culture was demonstrated by preparing a mixed culture of female and male cells that were respectively 'young' and 'old'. Unmixed cultures of each served as experimental controls. After 30 population doublings only the female cells remained in the mixed culture and only the female control cells were still dividing. The male cells had exhausted their population-doubling potential and had stopped dividing.

2. A cell-fusion experiment helped to determine the location of the clock that controls cellular aging. This was made possible by removing nuclei from cells and reinserting them into other cells. The cells without nuclei can survive for a short time. Nuclei from young cells (10 population doublings) were inserted into old cells (30 population doublings) and vice versa. In every case the nucleus was able to divide 50 times (10 + 40) (30 + 20) (10 + 40) (30 + 20) and no longer. The cell without nucleus did not influence the number of divisions. Hence, the clock is located in the chromosomes of the nucleus and not in the rest of the cell.

subsequently divide 40 times. The final result was that the total number of population doublings was always 50. Hayflick remarked that: "It is as if the cells have a built-in clock that counts divisions".

The question then arose as to whether this clock was present in the cytoplasm or the nucleus. Nuclear transplantation experiments gave the unequivocal answer that the clock that determined the ability to count cell division was in the nucleus (Fig. 4.14). Hence the cell memory turns out to be a chromosome memory. This means that the chromosome has the ability to count its DNA replications, and that it remembers how many times it has replicated.

Memory and time cannot be easily separated but "Time itself is as difficult to define as it is intuitively obvious" (von Baeyer, 1992)

Fig. 4.15   The connection between memory and time.

The Spanish artist Salvador Dali had close contact with scientists, especially physicists and biochemists. Time is most elusive and is one of the concepts in science that defies definition. Dali better than any one else expressed this situation in his classical painting called "The Persistence of Memory" in which watches melt down or are being devoured by ants as though they were dead matter (seen on the extreme left).

and its nature remains elusive since time cannot be properly apprehended (Hawking and Penrose, 1996). Few have dared to concretize time and memory on a canvas, but Dali, in his painting "The Persistence of Memory" (1931), combined the two concepts and expressed them in a most original way (Fig. 4.15).

96

When King Louis XV of France Was Going to be Married, the Princess in Question Could Give Birth to Rabbits

Louis XIV (1638–1715) became king at the age of five. He reigned in splendour for 72 years endowing France with a legacy of science, art, letters and technology.

The palace of Versailles, that he built near Paris, was so impressive in its architecture and decorations, that it became a model for royal residences across Europe from Lisbon to Stockholm. The intellectual elite of his time included scientists such as the mathematician Blaise Pascal (1623–1662), astronomer Pierre Gassendi (1592–1655) and physicist and philosopher René Descartes (1596–1650). Leading writers were: Jean Baptiste Molière (1622–1673), François de la Rochefoucauld (1613–1680) and Jean de la Fontaine (1621–1695). Painting was represented by: Antoine Watteau (1684–1721) and Nicolas Poussin (1594–1665). The founder of the French opera Jean Baptiste Lully (1632–1687) had an equally outstanding colleague in the musician François Couperin (died 1733). This was an exceptional rainbow of talents.

The Sun King, as he was known, left no direct progeny. His sons and grandsons had died. The successor was the young Louis XV (1710–1774). When he reached 15 years of age a princess had to be found who would marry the new king. After long enquiries into the noble families of France, a princess was found, who due to her intelligence and beauty, was considered the proper match. There was, however, a serious difficulty. Her mother — it was said — had given birth sometimes to a child, and at other occasions to a rabbit. The princess was put aside.

Thus, in the most educated circles of Europe, in 1725, there was no reason to exclude the possibility that a woman could at times produce a different animal. It is noteworthy that until recently, scientists, and in particular geneticists, had no critical argument to dismiss this assumption.

During the first 70 years of genetics (1900–1970) there was a divorce between embryology and genetics. The impressive order found in embryonic development could not fit in a genetic interpretation based on selection and random mutation. Development remained a black box and was ignored because it seemed unaccessible to genetic analysis.

It was only with the discovery of the so called *homeobox* genes that the development of the fly *Drosophila* could be reduced to gene intervention. It soon turned out that the same set of genes were responsible for the development of humans, mice, birds and flowers. The critical discovery was that within the "box" the genes were ordered and that this order was related to the order present in the adult body. In other words, the genes which occurred at one end of this DNA sequence were responsible for the formation of the organism's head, whereas the genes that were located at the opposite end of the "box" influenced the posterior part of the animal (the tail in the case of the mouse). Still more dramatic was the finding that this sequence of events had been preserved in flies, mice and humans (Lawrence, 1992) (Fig. 3.6).

The new genetic evidence has been most valuable in opening up this area of research but it only explains part of the embryonic process. This is why at present, we still do not know what inhibits a woman from producing a rabbit, since a large number of genes are common to all mammals.

97

Why Should a Woman Not Produce a Mouse

Until a few years ago it was known that mice and humans had about 50% genes in common. The remaining genes were then considered responsible for creating in one case a human and in the other a mouse. But, the base sequencing of the human genome as well as that of the mouse revealed that their genetic similarity is 99% (Goldstein, 2001; Hudson *et al.*, 2001).

Humans and mice shared a common ancestor about 100 million years ago. Despite such a long evolutionary distance between the two species they are most similar genetically. (1) There are many genes which are common to humans and mice building large blocks of conserved segments. (2) Genes that have been kept side by side in one mammalian species also tend to be linked in others. (3) There has also been a conservation of the gene linear order within the segments.

The similarity is so striking that a survey of homologous genes in human and mouse chromosomes led to establish 183 conserved segments. This resulted in the building of a comparative map between the human (22 + X + Y chromosomes) and mouse (19 + X + Y chromosomes). Some chromosomes turned out to be nearly identical. Almost all human genes on chromosome 17 are found on mouse chromosome 11 and the same is the case between human 20 and mouse 2. The largest contiguous conserved segment in the human genome is on chromosome 4 which is similar to mouse

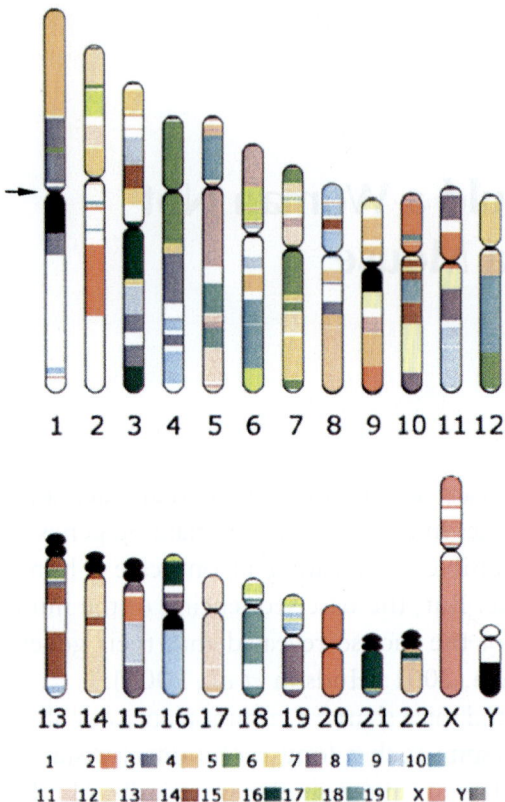

Fig. 4.16   The impressive similarity between mice and humans is revealed by conserved DNA sequences present in their chromosomes.

Human chromosomes (1 to 22 + X + Y) with segments, containing at least two genes, whose order is conserved in the mouse genome as colour blocks. Each colour corresponds to a particular mouse chromosome (1 to 19 + X + Y). The similarities between the two species are many and found in all chromosomes. Almost all human genes on chromosome 17 are found on mouse chromosome 11. Human chromosome 20 appears to be entirely similar to mouse chromosome 2. The largest conserved segment, which is apparently contiguous, is in human 4 and mouse 5. The human X chromosome (mainly responsible for sex determination) is also highly similar to the mouse X. The position of the centromere is indicated by an arrow. Centromeres, subcentromeric heterochromatin of chromosomes 1, 9 and 16, and the repetitive short arms of 13, 14, 15, 21 and 22 are in black.

chromosome 5 (International Human Genome Sequencing Consortium 2001, Mouse Genome Sequencing Consortium 2002) (Fig. 4.16).

These studies led to the conclusion that mice have about the same number of genes as humans, i.e. 30,000 and most of these genes have direct counterparts in humans. Even gene-poor regions of the chromosome, such as the heterochromatic areas, show extensive similarity between the two organisms.

The question then arises. Why are not mice more like us, or we not more like mice? The answer probably is to be searched for in the order of gene expression during the early stages of embryonic development, which remains so far a black box.

Hence, why does a woman not produce sometimes a mouse and at other times a child? Random genetic events would lead to such an unpredictable situation. So far it has not happened.

The worry that the noblesse of France had in the 1700s remains actual. One needs to await a deeper knowledge of the early steps in the rigid genetic road map to find out which molecular signals decide the canalization into either a mouse or a human.

References

Part IV

Darwin F. (1929) *Autobiography of Charles Darwin*. Watts and Co, London, UK.

Malthus TR. (1798) *An Essay on the Principle of Population as it Affects the Future Improvement of Society, with Remark on the Speculations of Mr. Godwin, M. Condorcet, and other writers*. Printed for J Johnson, London, UK.

Darwin C. (1859) *The Origin of Species by Means of Natural Selection or The Preservation of Favored Races in the Struggle for Life*. John Murray, London, UK.

Radl E. (1930) *History of Biological Theories*. Oxford University Press, Oxford, UK.

Hobsbawm E. (1998) *The Age of Capital 1848–1875*. Abacus, Little Brown Co, London, UK.

Lodge O. (1883) The ether and its functions. *Nature* **28**: 304–330.

David I, *et al.* (2002) *The Cambridge Dictionary of Scientists*. Cambridge University Press, Cambridge, UK.

Webster N. (1976) *Webster's New Twentieth Century Dictionary*. Collins World, USA.

Dobzhansky T. (1969) Selection. In: *Encyclopedia Britannica, Vol. 20*, pp. 181. William Benton, Chicago, USA.

Eigen M and Schuster P. (1979) *The Hypercycle. A Principle of Natural Self-organization*. Springer-Verlag, Berlin.

Frausto da Silva JJR and Williams RJP. (1994) *The Biological Chemistry of the Elements. The Inorganic Chemistry of Life.* Clarendon Press, Oxford.

de Duve C. (2005) The onset of selection. Natural selection started to drive evolution as soon as molecular replication became possible. *Nature* **433**: 581–582.

Cavalier-Smith T. (1985) Cell volume and the evolution of eukaryotic genome size. In: T Cavalier-Smith (ed), *The Evolution of Genome Size*, pp. 105–184. John Wiley and Sons, London.

Lima-de-Faria A. (2003) One *Hundred Years of Chromosome Research and What Remains to be Learned.* Kluwer Academic Publishers, Dordrecht, London (2003), Springer, NY (2004).

Lima-de-Faria A. (1983) *Molecular Evolution and Organization of the Chromosome.* Elsevier, Amsterdam, NY.

Lima-de-Faria A. (1988) *Evolution without Selection. Form and Function by Autoevolution.* Elsevier, Amsterdam, NY.

Lima-de-Faria A. (1997) The atomic basis of biological symmetry and periodicity. *BioSystems* **43**: 115–135.

Lima-de-Faria A. (2006) Is it time to rewrite all books in chemistry and physics? In: AM Pystin (ed), *Problems of Geology and Mineralogy*, pp. 76–82, *Geoprint, Russ Acad Sci*, Syktyvkar.

Britten RJ and Kohne DE. (1968) Repeated sequences in DNA. *Science* **161**: 529–540.

Birnstiel ML, *et al.* (1966) Localization of the ribosomal DNA complements in the nucleolar organizer region of *Xenopus laevis. Nat Cancer Inst Monogr* **23**: 431–447.

Brown DD, *et al.* (1971) Purification and some characteristics of 5S DNA from *Xenopus laevis. Proc Nat Acad Sci* **68**: 3175–3179.

International Human Genome Sequencing Consortium. (2001) Initial sequencing and analysis of the human genome. *Nature* **409**: 860–921.

Brown TA. (1999). *Genomes.* Bios Scientific Publishers, Oxford, UK.

Hood L. (2002) After the genome where should we go? In: M Yudell & R DeSalle (eds), *The Genomic Revolution*, pp. 64–73. Joseph Henry Press, Washington DC, USA.

Alberts B, *et al.* (1994) *Molecular Biology of the Cell.* Garland Publishing, NY, USA.

Brown TA. (2007). *Genomes 3.* Garland Science, NY, USA.

Tanaka H, *et al.* (2000) A ribonucleotide reductase gene involved in a p53-dependent cell-cycle checkpoint for DNA damage. *Nature* **404**: 42–49.

Lozano G, Elledge SJ. (2000). p53 sends nucleotides to repair DNA. *Nature* **404**: 24–25.

Begley TJ, Samson LD. (2003). A fix for RNA. *Nature* **421**: 795–796.

Aas PA, *et al.* (2003). Human and bacterial oxidative demethylases repair alkylation damage in both RNA and DNA. *Nature* **421**: 859–863.

Culbertson MR. (1999). RNA surveillance. Unforeseen consequences for gene expression, inherited genetic disorders and cancer. *TIG* **15**: 74–75.

Lykke-Andersen J, *et al.* (2001). Communication of the position of exon-exon junctions to the mRNA surveillance machinery by the protein RNPS1. *Science* **293**: 1836–1839.

Cole CN. (2001) Choreographing mRNA biogenesis. *Nat Genet* **29**: 6–7.

Stutz F, Izauarralde E. (2003) The interplay of nuclear mRNP assembly, mRNA surveillance and export. *Trends Cell Biol* **13**: 319–327.

Fasken MB, Corbett AH. (2005) Process or perish: quality control in mRNA biogenesis. *Nat Struct Mol Biol* **12**: 482–488.

Cullen BR. (2003) Nuclear RNA export. *J Cell Sci* **116**: 587–597.

Vinciguerra P, Stutz F. (2004). mRNA export: an assembly line from genes to nuclear pores. *Curr Opin Cell Biol* **16**: 285–292.

Becker WM, *et al.* (2003) *The World of the Cell, 5th Ed.* Benjamin Cummings, San Francisco, USA.

Lolle SJ, *et al.* (2005) Genome-wide non-mendelian inheritance of extra-genomic information in *Arabidopsis. Nature* **434**: 505–509.

Weigel D, Jürgens G. (2005) Hotheaded healer. *Nature* **434**: 443.

Tonegawa S, *et al.* (1978) Sequence of a mouse germ-line gene for a variable region of an immunoglobulin light chain. *Proc Natl Acad Sci* **75**: 1485–1489.

Strachan T, Read AP. (2004) *Human Molecular Genetics 3.* Garland Science, London and NY.

Pikaard CS. (2000) The epigenetics of nucleolar dominance. *TIG* **16**(11): 495–500.

Ferrari MR, *et al.* (2006) Nucleolar organizer activity and competition in Tricepiro Don René INTA, a synthetic forage crop. *Caryologia* **59**: 47–52.

Kleinjan DJ, van Heyningen V. (1998) Position effect in human genetic disease. *Hum Mol Genet* 7: 1611–1618.

Bell AC, *et al.* (2001) Insulators and boundaries: versatile regulatory elements in the eukaryotic genome. *Science* **291**: 447–450.

Novick RP. (1980). Plasmids. *Sci Am* **243**(6): 76–90.

Sheidai M., *et al.* (2006) Cytogenetic variability in several Canola (*Brassica napus*) cultivars. *Caryologia* **59**: 267–276.

Crow JF. (1979) Genes that violate Mendel's rules. *Sci Am* **240**: 104–113.

Ritossa F. (1973) Crossing-over between X and Y chromosomes during ribosomal DNA magnification in *D melanogaster. Proc Nat Acad Sci* **70**: 1950–1954.

Tartof KD. (1974) Unequal mitotic sister chromatid exchange as the mechanism of ribosomal RNA gene magnification. *Proc Nat Acad Sci* **71**: 1272–1276.

Donelson JE, Turner MJ. (1985) How the Trypanosome changes its coat. *Sci Am* **252**(2): 32–39.

Evans SK, Lundblad V. (2000). Positive and negative regulation of telomerase access to the telomere. *J Cell Sci* **113**: 3357–3364.

Cairns J. (1998) Mutation and cancer: the antecedents to our studies of adaptive mutation. *Genetics* **148**: 1433–1440.

Kuczmarski ER, Spudich JA. (1980) Phosphorylation of myosin heavy chain *Dictyostelium* regulates self assembly. *20th Ann Meeting Am Soc Cell Biol J Cell Biol* **87**: 227A.

de la Roche MA, *et al.* (2002) Signaling pathways regulating *Dictyostelium* myosin II. *J Muscle Res Cell Motility* **23**: 703–718.

Ginger RS, *et al.* (1998) A novel *Dictyostelium* cell surface protein important in both cell aggregation and cell sorting. *Development* **125**: 3343–3352.

Loomis WF, Insall RH. (1999) A cell for all seasons. *Nature* **401**: 440–441.

Gilbert SF. (2000) *Developmental Biology*. Sinauer Associates, Sunderland, MA, USA.

Edgar B, *et al.* (1994) MPF regulation during the embryonic cell cycles of *Drosophila*. *Genes Dev* **8**: 440–453.

Carrington C, Ambros V. (2003) Role of microRNAs in plant and animal development. *Science* **301**: 336–338.

Denli AM, Hannon GJ. (2003) RNAi: an ever-growing puzzle. *Trends Biochem Sci* **28**: 196–201.

Matzke M, Matzke AJM. (2003) RNAi extends its reach. *Science* **301**: 1060–1061.

Townes PL, Holtfreter J. (1955) Directed movements and selective adhesion of embryonic amphibian cells. *J Exp Zool* **128**: 53–120.

Moscona AA. (1952) Cell suspension from organ rudiments of chick embryos. *Exp Cell Res* **3**: 535–539.

Godt D, Tepass U. (1998) *Drosophila* oocyte localization is mediated by differential cadherin-based adhesion. *Nature* **395**: 387–391.

Gumbiner BM. (2005) Regulation of cadherin-mediated adhesion in morphogenesis. *Nat Rev Mol Cell Biol* **6**: 622–634.

Halbleib JM, Nelson WJ. (2006) Cadherines in development: cell adhesion, sorting, and tissue morphogenesis. *Genes Dev* **20**: 3199–3214.

Henry JJ, Raff RA. (1990) Evolutionary change in the process of dorsoventral axis determination in the direct developing sea urchin, *Heliocidaris erythrogramma*. *Dev Biol* **141**: 55–69.

Henry JJ, *et al.* (1989) Early inductive interactions are involved in restricting cell fates of mesomeres in sea urchin embryos. *Dev Biol* **136**: 140–153.

Khaner O, Wilt F. (1991) Interactions of different vegetal cells with mesomeres during early stages of the sea urchin development. *Development* **112**: 881–890.

Ganz T. (2003) Defensins: antimicrobial peptides of innate immunity. *Nat Rev Immunol* **3**: 710–720.

Smith LC. (2001) The complement system in sea urchins. *Adv Exp Med Biol* **484**: 363–372.

Hoffmann JA, *et al.* (1999) Phylogenetic perspectives in innate immunity. *Science* **284**: 1313–1318.

Agrawal A. (2000) Transposition and evolution of antigen-specific immunity. *Science* **290**: 1715–1716.

Janeway CA Jr, *et al.* (2005) *Immunobiology, 6th Ed.* Garland Science, NY, USA.

Brenner S. (1974) The genetics of *Caenorhabditis elegans. Genetics* **77**: 71–94.

Hengartner MO, *et al.* (1992) *Caenorhabditis elegans* gene *ced-9* protects from programmed cell death. *Nature* **356**: 494–499.

Adams JM, Cory S. (1998) The Bel-2 protein family: arbiters of cell survival. *Science* **281**: 1322–1326.

Gurdon JB, *et al.* (1994) Activin signalling and response to a morphogen gradient. *Nature* **371**: 487–492.

Luigi B, *et al.* (2006) Programmed cell death in the nucellus of *Tillandsia* (Bromeliaceae). *Caryologia* **59**: 334–339.

Kirk DL. (1999) Evolution of multicellularity in the Volvocine lineage. *Curr Opin Plant Biol* **2**: 496–501.

Bassnett S. (2002) Lens organelle degradation. *Exp Eye Res* **74**: 1–6.

Dahm R. (2004) Dying to see. *Sci Am* **2004**: 52–59.

Janeway CA Jr, *et al.* (2005) *Immunobiology, 6th Ed.* Garland Science, NY, USA.

Turner BM. (2002) Cellular memory and the histone code. *Cell* **111**: 285–291.

Hayflick L. (1980) The cell biology of human aging. *Sci Am* **242**(1): 42–49.

von Baeyer HC. (1992) *Taming the Atom.* Random House, NY, USA.

Hawking SW, Penrose R. (1996) The nature of space and time. *Sci Am* **275**(1): 44–49.

Lawrence PA. (1992) *The Making of a Fly.* Blackwell Scientific Publications, Oxford, UK.

Goldstein JL. (2001) Laskers for 2001: knockout mice and test-tube babies. *Nat Med* **7**(10): 1079–1080.

Hudson TJ, *et al.* (2001) A radiation hybrid map of mouse genes. *Nat Genet* **29**: 201–205.

International Human Genome Sequencing Consortium. (2001) Initial sequencing and analysis of the human genome. *Nature* **409**: 860–921.

Mouse Genome Sequencing Consortium. (2002) Initial sequencing and comparative analysis of the mouse genome. *Nature* **420**: 520–562.

Gilbert SF. (2000) *Developmental Biology*. Sinauer Associates, Sunderland, MA, USA (cited on Fig. 4.11).

Sources of Illustrations

Part IV

4.1   Hobsbawm E. (1998) *The Age of Capital, 1848-1875*. Abacus, Little Brown, London, UK. (Plate 1, page 144).

4.2   Dudesq AJ, Rudel J. (1968) *Collection d'Histoire, 1789-1848*. Bordas, Paris, France (Fig. 508, page 523).

4.3   Brown TA. (2007) *Genomes 3*. Garland Science, NY, USA (Fig. 16.20, page 524).

4.4   Brown TA. (2007) *Genomes 3*. Garland Science, NY, USA (Fig. 16.29, page 532).

4.5   Fasken MB, Corbett AH. (2005) Process or perish: quality control in mRNA biogenesis. *Nat Struct Mol Biol* **12**: 482–488 (Fig. 1, page 483).

4.6   Cole CN. (2001) Choreographing mRNA biogenesis. *Nat Genet* **29**: 6–7 (Fig. on page 7).

4.7   Begley TJ, Samson LD. (2003). A fix for RNA. *Nature* **421**: 795–796 (Fig. 1, page 795).

4.8   Weigel D, Jürgens G. (2005) Hotheaded healer. *Nature* **434**: 443 (Fig. 1, page 443).

4.9   Gilbert SF. (2000) *Developmental Biology*. Sinauer Associates Inc Publishers, Sunderland, MA, USA (Fig. 3.11, page 57).

4.10   Halbleib JM, Nelson WJ. (2006) Cadherines in development: cell adhesion, sorting, and tissue morphogenesis. *Genes Dev* **20**: 3199–3214 (Fig. 4, page 3203).

4.11   Gilbert SF. (2000) *Developmental Biology.* Sinauer Associates Inc. Publishers, Sunderland, MA, USA (Fig. 8.7, page 229).

4.12   (1)   Raff RA. (1996). *The Shape of Life.* University of Chicago Press, Chicago, USA (Fig. 6.4, page 194) (From Haeckel, E (1874) Anthropogenie, oder Entwickelungsgeschichte des Menschen. Leipzig, Engelmann, Germany).

         (2)   Gilbert SF. (2000). *Developmental Biology.* Sinauer Associates Inc Publishers, Sunderland, MA, USA. (Fig. 118, page 19).

4.13. Lodish H, *et al.* (1995) *Molecular Cell Biology.* Scientific American Books, WH Freeman and Co, NY, USA. (Fig. 27.10, page 1305) (After Hood L, *et al.* (1984) *Immunology, 2nd ed.* Benjamin, USA, page 11).

4.14   (1) and (2) Hayflick L. (1980) The cell biology of human aging. *Sci Am* **242**(1): 42–49. Adapted from: Fig. Limit of replication, page 46; Fig. Cell-fusion experiment, page 48. In: Lima-de-Faria A. (1983) *Molecular Evolution and Organization of the Chromosome.* Elsevier, London (Fig. 42.10, page 1044; Fig. 42.11, page 1044).

4.15   Dali S. (1931) *The Persistence of Memory.* The Museum of Modern Art, NY. Honour H, Fleming J. (2002). *A World History of Art* (Fig. 20.17, page 822).

4.16   International Human Genome Sequencing Consortium. (2001) Initial sequencing and analysis of the human genome. *Nature* **409**: 860–921 (Fig. 46, page 910).

Who Cares for Magnetism

Who Cares for Magnetism

98

Magnetism and Electricity are Two Manifestations of the Same Phenomenon

Chromosomes are not magnets, yet they share some of their properties.

Physicists in an effort to understand the behavior of matter and energy distinguish between four separate forces which prevail in the universe. It may be recalled that they are: the electromagnetic force, the strong force, the weak force and gravitation (Fig. 2.1). The first force has a double name because electricity and magnetism can be converted into one another. As a consequence these two original forces were unified into a single one.

Previously, we dealt with the behavior of the chromosome in relation to gravity. Let us now examine its relationship with magnetism (Fig. 5.1).

There is nothing obscure about such an inquiry. We tend to forget that: 1) Cells, including those of the human brain, send electrical signals regularly. 2) Magnetite particles occur in the brain of animals. 3) Electricity and magnetism have been shown, long ago, to represent two manifestations of the same phenomenon. They are two sides of the same coin.

As early as 1820, the Danish physicist Hans C. Oersted discovered that an electrical current produces a magnetic field. This relationship was demonstrated by a simple experiment. He placed a

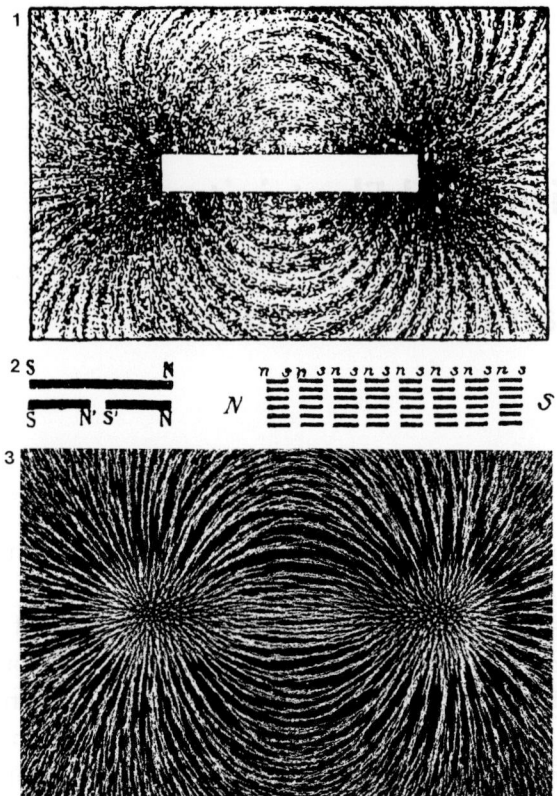

Fig. 5.1   The formation of a magnetic field is concretized by the distribution of iron particles.

1. Photograph of the magnetic field created by a mineral such as magnetite which is known to attract iron. A paper was first covered with small particles of iron and then a rod of magnetite was located in their vicinity. The small particles were found to orient themselves according to a field pattern directed by the two poles.

2. A magnet with its South (S) and North (N) poles. When the magnet was divided into two pieces, this process did not result in the isolation of one of its poles, but each piece acquired its own South and North pole, i.e., the magnetic field was reshaped and maintained, irrespective of change in size. If many small magnets were aligned, and put together with all their North poles facing in the same direction, a large magnet was reconstituted in which the magnetic moment was equal to the sum of the moments of the small magnets (the magnetic moment is the force multiplied by the distance between the poles). That is, if there was an increase in size, then the field adjusted accordingly.

3. Photograph of iron filings scattered on a card above a bar magnet revealing the pattern of the magnetic field around the magnet.

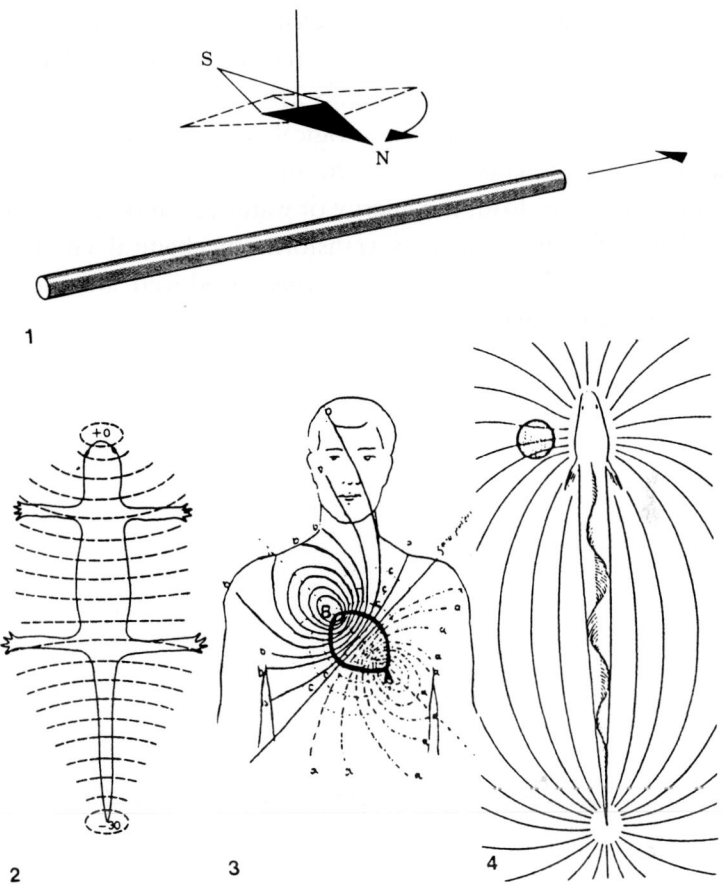

Fig. 5.2 Electricity and magnetism are two manifestations of the same phenomenon.

1. The Danish physicist Hans Oersted (1777–1851) astonished his students in 1820 when he showed that a magnetic compass needle, that was parallel to a wire, moved to a perpendicular position when an electrical current was sent through the wire. The electrical current induced a magnetic field around it that obliged the needle to move.

2. Electric field produced by adult newt determined by two recording electrodes at the tip of the nose and tail.

3. Electricity produced by the human heart during contraction as recorded in an electrocardiogram.

4. Current flowing between the electric organ in the tail of a fish and receptor pores in the head.

magnetic compass needle directly below a wire. When the electric current was sent through the wire, the needle moved due to the induction of a magnetic field (Fig. 5.2).

His discovery led to a technological surge that resulted in the creation of the dynamo. This is the machine used everywhere to produce electricity from the energy of waterfalls and other sources. The dynamo's large magnets transform mechanical energy into electricity, the electrical current being produced by the rapid movement of the magnets.

Bacteria, Bees, and Pigeons Orient According to the Magnetic Field

Experiments are revealing the effect of the Earth's magnetic field on living organisms and the mechanism involved.

Bacteria from marine sediments rapidly migrate in response to the local geomagnetic field. The direction of movement is immediately changed when small magnets are moved about in the vicinity of the bacteria. They react to fields as weak as 0.5 gauss. The orientation of the bacteria is due to the presence in their cells of chains of crystals of magnetite (Fig. 5.3). By changing the position of the magnetic fields it was ascertained that the bacterial movement, which takes place with the help of flagella, was directed by the Earth's magnetic field (Blakemore, 1975).

Another group of bacteria from freshwater sediments were found to be magnetotactic. Their cells contained 100×150 nm iron crystals. These cells had an average of 22 intracellular crystals and up to 1.5% of their dry weight was iron. A spectroscopic analysis disclosed that the cellular iron was in the form of magnetite (Frankel *et al.*, 1979; Lins *et al.*, 2006) (Fig. 5.3).

Bees show several effects in magnetic fields. (1) When a swarm of bees is deprived of a means of orientation they build their honeycomb in the same magnetic direction as in the parent hive. (2) When no other stimuli are involved, the bees seem to set their biological clock after the regular daily variations in the magnetic field of the Earth. (3) When an abnormally strong magnetic field is

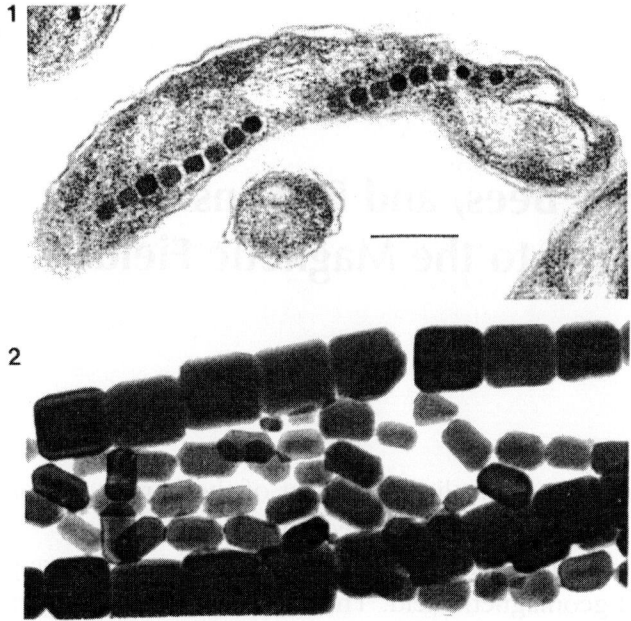

Fig. 5.3 Bacteria swim, oriented by magnetic fields, due to the presence in their cells of magnetite crystals.

1. Electronmicrograph of a section of freshwater magnetotactic bacteria showing chains of crystals of magnetite. The bar is 250 nm.

2. Electronmicrograph of isolated chains of membrane enclosed iron crystals present in magnetotactic bacteria. These crystals are of magnetite (Fe_3O_4) and occur in two size classes up to 250 and 120 nm. The chains orient close to the axis of mobility of the bacterial cell building a dipole that causes the cell to remain oriented along the local magnetic field as the bacterium swims.

applied their rhythm is disturbed. That bees have magnetic remanence was shown by applying a strong field of about 700 gauss (the Earth's field is circa 0.5 gauss) to bees. The magnetic material was located exclusively in the frontal region of the abdomen. Magnetite was identified in these tissues as the primary magnetic element (Gould *et al.*, 1978).

Homing pigeons have both a "map" and "compass sense" since they find their home location after being released at an unfamiliar site. It is well established that on sunny days they orient by the sun, but they also find their way in cloudy weather. When small magnets are attached to their heads the ability to orient under cloudy conditions is disrupted. This indicates that magnetic field information is used for orientation. A search was then made for stable and superparamagnetic domains in pigeons with the help of a magnetometer. In every pigeon tested magnetic material was present, located unilaterally in a tissue intimately associated with the skull. The analysis of the crystals in the pigeon tissue revealed that they consisted of magnetite (Walcott *et al.*, 1979).

100

Cells Generate Electricity and Magnetism

Crystals already generate electricity. When pressure is exerted at the ends of the polar axis of a crystal, a flow of electrons takes place that charges the crystal with negative electricity at one end and positive electricity at the other. The phenomenon was described in quartz by Pierre Curie (the co-discoverer of radioactivity) in 1881, but it was only after 1921 that it became of general application in radios and watches. As pressure creates electricity, conversely an electric current leads to a deformation of the crystal. The result is that a slice of quartz subjected to an alternating current vibrates, and as such, becomes the essential component of a quartz watch (Pitt, 1988).

Every cell generates electrical currents. Muscle and nerve cells have been used in the experimental demonstration of the production of electricity since it is most evident in these tissues. A nerve cell is a tiny battery capable of generating an electrical impulse that travels along its fibers for several hundred centimeters without becoming attenuated (Loewy and Siekevitz, 1970). Moreover, many animal organs produce powerful electrical fields. These may extend outside the organ, as in the case of the human heart, its current being recorded routinely in hospitals in the form of electrocardiograms. The electricity may reach outside the organism's body as in the electric eel and the electric ray, with strengths of about 200 volts (Romer and Parsons, 1978) (Fig. 5.2).

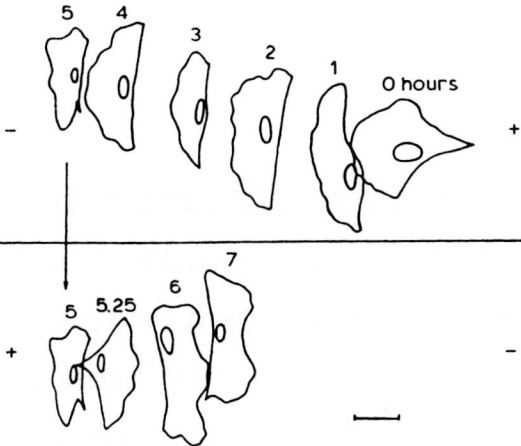

Fig. 5.4   Cells moving in electrical fields.

Directional movement of a quail cell in an electrical field. The cell migrates toward the negative pole for 5 hours. When the current was reversed, at exactly this time, the cell changed its direction, moving to the new cathode.

The transformation of electricity into magnetism has been recorded in human cells. The electric currents sent by the human heart produce magnetic fields that have been measured with a magnetocardiograph.

Water and living cells which consist mainly of water, are known to be weakly magnetic. They become magnetized in the direction opposite to an applied magnetic field (diamagnetism) (Schneider, 1999).

Electric currents occur during embryonic cell movements. The motility of embryonic cells of quail showed a striking sensitivity to small, steady electric fields. The cells responded in three ways: (1) They oriented with their long axes perpendicular to the field lines in fields of 150 mV/mm to 600 mV/mm. (2) The cells migrated to the cathode and this movement could be reversed by reversing the electric current. (3) One hour after applying a field of 400 mV/mm the cells elongated and aligned perpendicularly to the field lines. The average speed of cell movement is circa 1μm/min (Nuccitelli and Erickson, 1983) (Fig. 5.4).

Let us now find out how magnets behave and how the properties of chromosomes compare with their properties.

101

When Magnets Are Divided into Minor Pieces, Each Separate Unit Continues to Behave as a Magnet, Acquiring New North and South Poles, But Small Magnets Can Also Rebuild Large Ones

Our planet possesses a strong magnetic field that has varied throughout the ages. The natural magnetism found in the mineral magnetite (Fe_3O_4) is considered to have arisen during the cooling process that occurred at the time of the earth's formation. This mineral attracts iron displaying the following properties:

(1) When a magnet is placed in a vial containing small iron particles, these attach mainly to its extremities, called poles. They are designated south and north, since they orient to these regions of the earth. The same terrestrial magnetism obliges the needle of the mariner's compass to orient to the north pole.

(2) A magnet cut into smaller pieces would be expected to show isolation of the north from the south pole. This is not what occurs. Instead two complete magnets are formed, each with a south and a north pole. If division continues, the result is always the same. New complete magnets arise; this pattern is maintained irrespective of size (Fig. 5.1).

(3) The reverse experiment can be carried out. If small magnets are put together and aligned in such a way that the north pole of one coincides with the south pole of the next, then a large magnet is built, whose power equals the sum of the small magnets. Hence, the functional pattern is maintained irrespective of the decrease or increase in size, once specific conditions are met (Fig. 5.1).

(4) The magnetic field is easily visualized by spreading small particles of iron over a sheet of paper and placing under it a magnetized bar. At once the iron particles dispose themselves around the poles, following the lines of force that extend from pole to pole.

(5) Magnetism is considered to be due to unpaired electrons in certain atom orbitals as well as changes in atomic spin (Bloss, 1971).

The remarkable property of magnets is that they can be disassembled into minor units which retain the properties of the original larger whole and at the same time the minor units can reassemble building a larger body which retains the properties of the smaller components.

This capacity of readjustment of the field pattern, irrespective of size, turns out to have its counterpart in the behavior of cells and chromosomes.

102

When Fertilized Eggs Are Divided into Separate Cells, Each Cell Acquires the Properties of the Initial Egg Giving Rise to Separate Embryos

Most people know that a fertilized egg gives rise to an embryo. This occurs by successive divisions of the original cell leading to the formation of a new organism. What is less known is that by a simple experiment, when the fertilized egg has divided into four cells, each of these cells can be isolated and grown independently. As early as 1892, Hans Driesch separated sea urchin cells from each other. He first removed the fertilization envelope that surrounded the embryo at the four cell stage. Then he submitted it to vigorous shaking. To Driesch's surprise, each of the isolated cells produced entire sea urchin larvae (Fig. 5.5).

The result was startling and has been repeated many times since then. Each isolated cell of the quartet was able to regulate its development so as to produce a complete organism. If a magnet is divided into four parts, each of the new minor parts behaves functionally as the original large magnet due to an internal regulation of its atomic properties. In the egg a comparable situation occurs, each of the minor four parts is able to behave functionally as the original large egg. The mechanism involved in this self-regulation is not any mystic force, but it is at present well established.

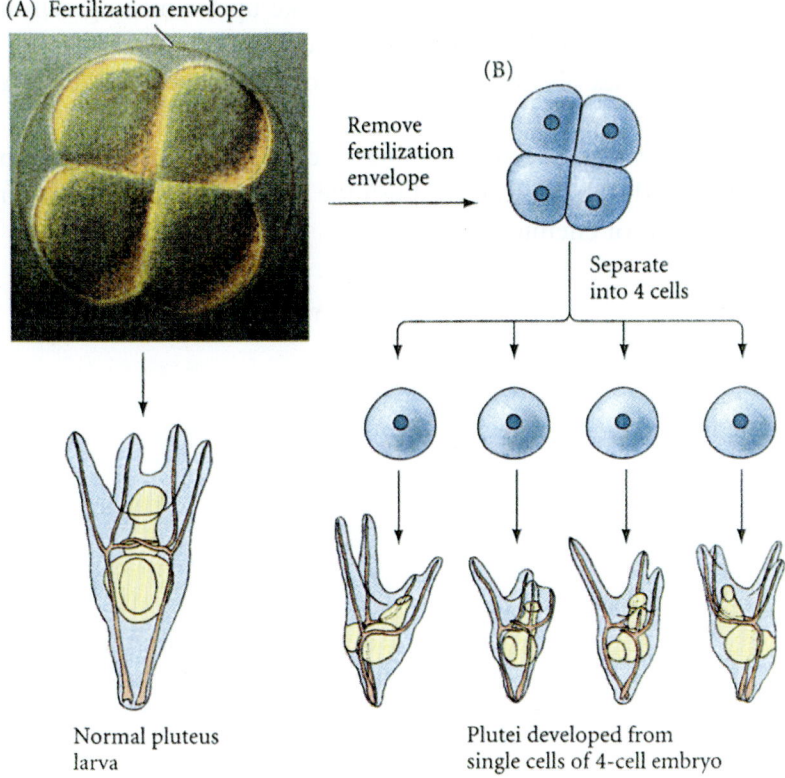

Fig. 5.5   Each single cell is able to form a perfect organism.

Hans Driesch's demonstration (1892) of regulative development in sea urchins. (A) An intact 4-cell sea urchin embryo generates a normal pluteus larva. (B) When one removes the 4-cell embryo from its fertilization envelope and isolates each of the four cells, each cell can form a smaller, but normal, pluteus larva. (Photograph courtesy of G. Watchmaker.)

During embryonic development cell fates are determined by soluble molecules secreted at a distance from the target cells. One such molecule is the protein activin which builds a concentration gradient within the embryo causing different gene expression in different cells as it attaches to receptors on the cell's membrane (Fukui and Asashima, 1994).

Other experiments have disclosed that the cells of the embryonic field (as it is called by embryologists) can regulate their fates, i.e. their functional patterns, to make up for missing cells in the field (Huxley and de Beer, 1934; de Robertis *et al.*, 1991). Foreign cells may also be integrated into the organized field without disturbing the pattern.

In other words, if cells are subtracted or added to the embryo, this one has the ability to reestablish the original functional pattern with the help of chemical signals.

103

Separate Embryos Which Are Fused Result in a Single Normal Organism

Like in magnets, the reverse process of aggregation of cells and embryos into large units resulted in the recreation of the pattern of

Fig. 5.6 A single normal organism is produced from several separate embryos.

A sheep-goat chimaera produced by combining one 8-cell sheep embryo with three 8-cell goat embryos. The animal is about 1 year old and behaves normally. Its chimaeric nature is confirmed by the following characters: (1) a mosaic distribution of hairy and wooly areas in its coat, (2) goat-like horns twisted like sheep horns and (3) blood containing sheep and goat red blood cells.

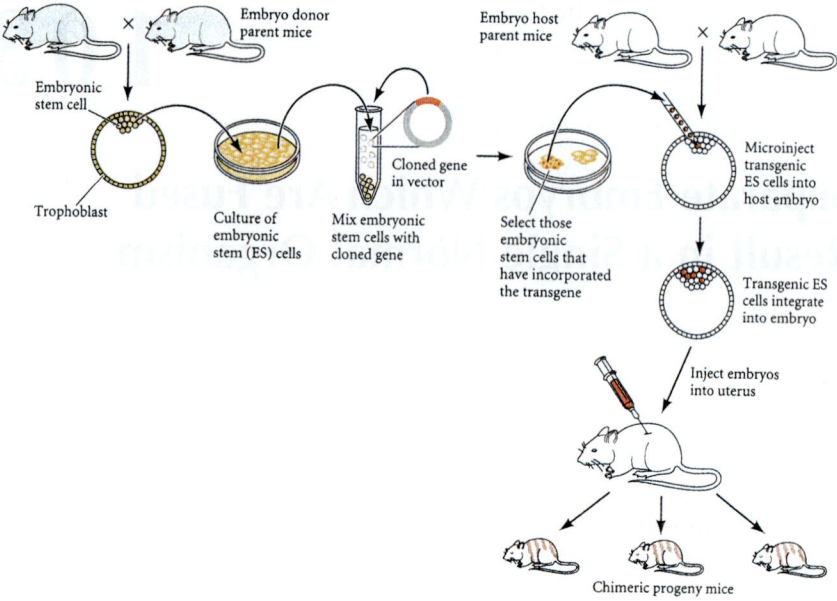

Fig. 5.7   Cells reshape themselves into a single embryo forming a normal mouse.

In the production of transgenic mice embryonic stem cells from one mouse are cultured and their genome altered by the addition of a cloned gene. These transgenic cells are selected and then injected into the early stages of a host mouse embryo. Here, the transgenic embryonic stem cells integrate with the host's embryonic cells. The embryo is placed into the uterus of a pregnant mouse and develops into a normal chimaeric mouse. The added gene (the transgene) can be from any animal source.

the minor components. This finding was confirmed by several experiments.

An adult animal, with the traits of a goat and a sheep (a sheep-goat chimaera) was produced by combining one 8-cell sheep embryo with three 8-cell goat embryos. These reorganized themselves into a single embryo which developed into a normal individual (Fehilly *et al.*, 1984) (Fig. 5.6).

A similar experiment had been previously carried out in mice. In this case two embryos, from mothers with different genetic constitutions, were fused. The cells reshaped themselves into a single embryo that led to the formation of a normal mouse. The addition of a foreign group of cells to another embryo led to a similar result. They were incorporated and integrated into the original pattern producing a single individual (Mintz and Illmensee, 1975; Mintz, 1978) (Fig. 5.7).

104

When Chromosomes Are Divided into Minor Pieces, Each Separate Unit Continues to Behave as an Independent Chromosome by Incorporating or Creating New Telomeres and Centromeres

Due to various cellular disturbances, or irradiation with X-rays, chromosome fragments arose frequently during cell division. The question was then asked: Did these fragments, of many different sizes, survive equally well as the original chromosomes? Already, during the early days of genetics the Russian cytologist M. Nawashin showed that, for a fragment to survive, it had to possess a centromere, the region that allowed its regular movement to the poles during cell division. Soon the American geneticist J.H. Muller proved that the fragments also had to be capped at each end by a specific region called telomere, if they were to be maintained by the cell.

For several decades it was thought that these two important regions had to be acquired from other chromosomes by accidental processes if the fragment was to survive. However, during the last 15 years a different picture has emerged.

(1) Chromosomes contain silent secondary centromeres and telomeres that may be activated when they are broken into

minor pieces. They do not necessarily need to incorporate foreign ones. They can reorganize themselves and become functional. The American cytologist Barbara McClintock showed that chromosomes of maize could loose their telomeres and continued to behave normally. In humans once the centromere ceased to function, previously silent centromeres took over the function reestablishing the normal chromosome activity (Niebuhr and Skovby, 1977; Merry *et al.*, 1985). Significant is that, following chromosome breakage, new telomeres may be formed in the chromosomes of *Ascaris* (a nematode worm) (Müller *et al.*, 1991).

(2)   Both centromeres and telomeres are normally compound structures. Two new chromosomes may be formed from a single one, each piece having half, or even less than half, of the original centromere and behave quite normally during cell division.

(3)   Even more surprising, centromeres can suddenly behave as telomeres. Centromeres, which usually are located in the median regions of chromosomes, can become distally located. The centromere then becomes the natural cap behaving as a telomere. This is not an exceptional situation but it is the general feature in several animal groups such as grasshoppers and lizards. In these animals most chromosomes of the normal complement have a strictly terminal centromere which functions both as a telomere and as a centromere (White, 1973).

Hence, the chromosome has the ability to reorganize itself, creating novel functions in some of its regions. This permits a fragment to reshape itself acquiring the properties of the original unit.

105

The Same Chromosomes May Disassemble and Reassemble Maintaining Their Genetic Properties — A Deer Species May be Formed with 35 or Only 3 Chromosomes

An unexpected event is that the same type of organism, be it a protozoan or a mammal, can be found in nature having extreme numbers of chromosomes. These have arisen by breakage of large chromosomes into smaller units followed by their dispersal and individualization into independent chromosomes. The reverse situation has occurred as well, the smaller chromosomes reuniting to build again larger chromosomes.

The chromosome number can reach an astonishing variation. Only 2, or as many as 500 chromosomes can occur within the same group. Yet the basic genetic solution is the same.

In higher organisms, most body cells, i.e. the so called somatic cells, contain double the chromosome number of the sexual cells.

The deer (Cervidae) show the largest variation in chromosome number found within the same mammalian family. The somatic cells of red deer (*Cervus elaphus*) have 68 and those of the reindeer

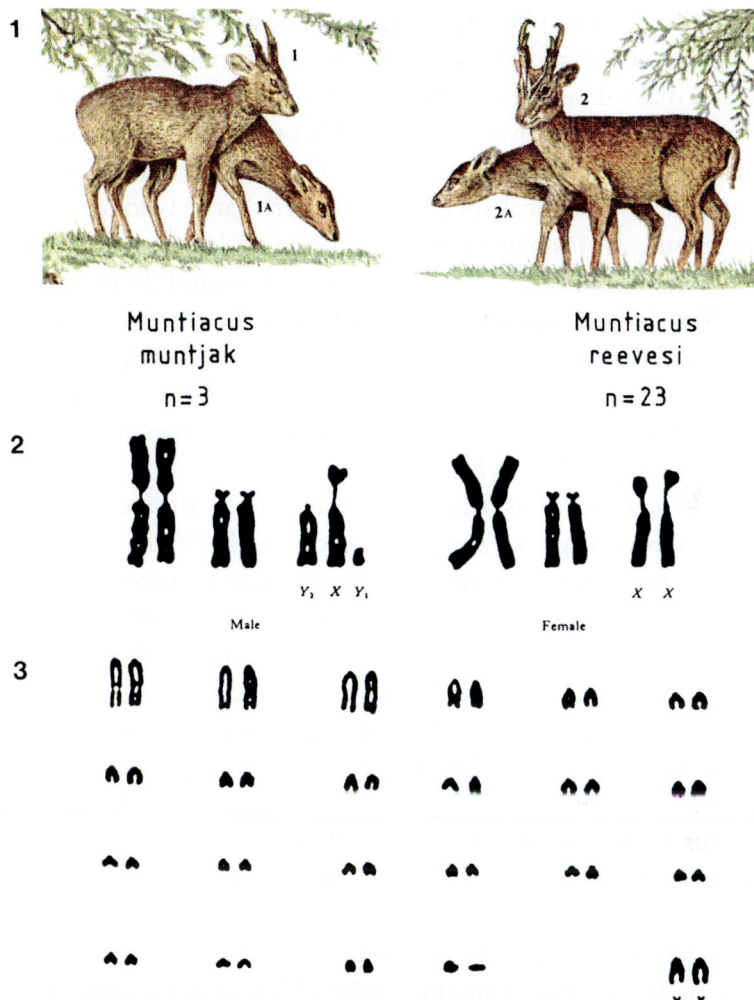

Fig. 5.8 Nearly identical animals are produced with 46 or 6 chromosomes.

1. *Muntiacus muntjak*, female n = 3 (1A), male n = 3 + y$_2$ (1). *Muntiacus reevesi* n = 23, female (2A), male (2). The females are practically identical despite the extreme variation in chromosome number, and the males have horns that are only slightly different. *n* is the chromosome number found in germ cells. The other cells of the body have, in most organisms, the double number of chromosomes (somatic number).

2. The chromosomes of *Muntiacus muntjak* (male 7, female 6).

3. The 46 chromosomes of *Muntiacus reevesi* (female).

(*Rangifer tarandus*) 70 chromosomes. Most body cells of the deer *Muntiacus muntjak* have only 6 chromosomes in the female and 7 in the male. This means that the germ cells possess only 3 and 4 chromosomes respectively (the extra chromosome in the male is an additional minor sex chromosome). Cytologists were bewildered when a nearly identical species, so similar that it belonged to the same genus, *Muntiacus reevesi*, was found with the same chromosome number as humans, 46 in somatic tissues and 23 in the germ cells. Not only the number is the same, but the chromosome complement of *M. reevesi* consists of groups of large, medium size and small chromosomes as found in the human genome. The large size of the sex chromosomes in the females of both species is another common feature (Fig. 5.8).

Most significant is that in the muntjacs the females with 3 chromosomes are so similar to those with 23 that zoologists can only sort them out after analyzing their chromosome numbers. The body of the males is also identical, only the horns differ slightly (Fig. 5.8). Although the structural genes, and other DNA sequences, have been dispersed, or concentrated, into a highly diverse number of units, the final result of all genetic functions has led to the production of a nearly identical organism.

Several laboratories have, in the last years, contributed to elucidate this situation. The DNAs of these species were extracted, cleaved with restriction enzymes and submitted to DNA-DNA hybridization. The experiments disclosed that the three chromosomes of *M. muntjak* contained many DNA sequences which were homologous to the other deer species with a chromosome number 10 times higher, such as red deer (Lima-de-Faria *et al.*, 1986). Other techniques involving "chromosome libraries", composite DNA probes, labelling with fluorescent agents, and centromere specific antiserum, disclosed other homologies between the different deer species (Brinkley *et al.*, 1984; Goldoni *et al.*, 1984; Scherthan *et al.*, 1994). For instance, the centromeres of the

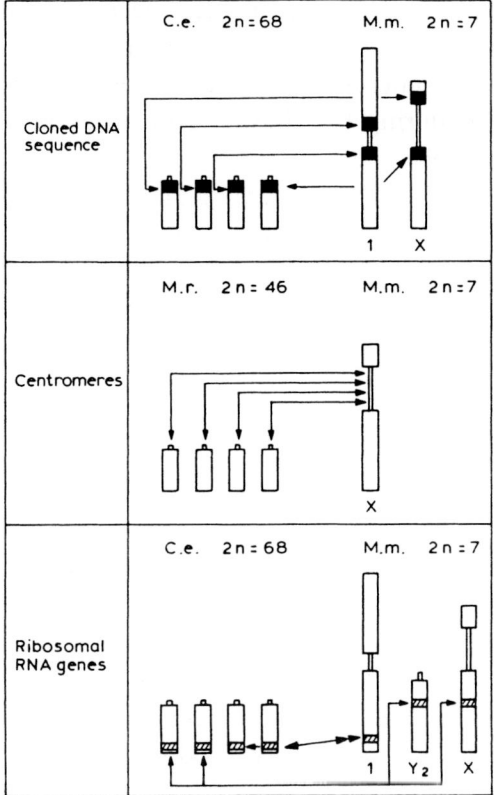

Fig. 5.9   Diagram summarizing the evidence in deer species demonstrating the existence of order in chromosome reorganization.

Cloned DNA sequences, centromeres and ribosomal RNA genes maintain their main chromosome territory, despite drastic chromosome rearrangements. For the sake of simplicity only 4 chromosomes are represented out of 68 and 46. For the same reason only some of the chromosomes of *M. muntjak* are represented. The arrows indicate that the rearrangements may occur in both directions, since they may take place from a high to a low chromosome number but also in the reverse direction as is known to happen in chromosome evolution. (C.e. = red deer, M.m. = *M. muntjak*, M.r. = *M. reevesi*). *2n* is the somatic chromosome number found in most cells of the body with exception of the sexual cells. During the reshaping of the chromosomes: (1) Cloned DNA sequences maintained their position near centromeres. (2) Centromeres moved to a strict centromere position piling themselves on top of each other. (3) Ribosomal RNA genes, which tend to occupy a terminal position, maintained this location in one case but this deviated in two other instances.

46 chromosomes of *M. reevesi* had not been dispersed at random but had become aligned and packed, alongside each other, building gigantic centromeres in the 3 chromosomes of *M. muntjak*. Order prevailed during the drastic reorganization of these chromosomes (Fig. 5.9).

106

Ants May be Produced Using a Single Chromosome But Also 94

In insects such as the ants, the solution is extreme. Almost 500 species of this animal group have had their chromosomes counted (Taber and Cokendolpher, 1988). The ant *Myrmecia pilosula*, has closely related species with 9, 10, 16, 24, 30, 31 and 32 chromosomes (somatic cells), but siblings "currently indistinguishable morphologically" from these have turned out to have only 2 chromosomes in the females and one single chromosome in the males. Like in honey bees, the females develop from fertilized eggs, but the males are produced without the eggs being fertilized, the result being that they have only half of the chromosomes, which in this case is a single one (Crosland and Crozier, 1986) (Fig. 5.10).

There are other ant species that may have 94 chromosomes in their somatic tissues (Imai *et al.*, 1990). Hence the variation is from 1 to 94, and yet the final genetic product is still an ant.

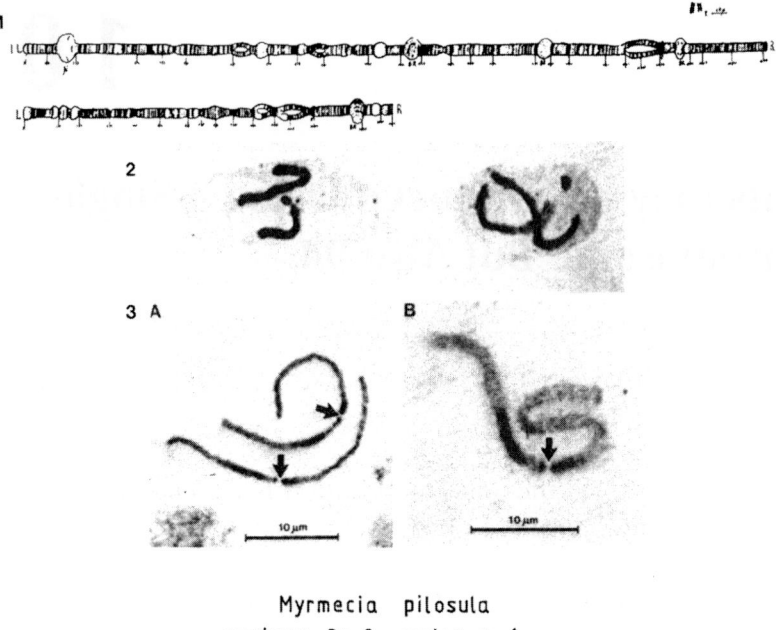

Myrmecia   pilosula
workers 2n=2   males n=1

Fig. 5.10   A single chromosome can produce a perfect animal.

1. The two large chromosomes of the insect *Cricotopus silvestris* in salivary glands and in another tissue at mitosis exhibiting their minute size (upper right) at the side of a 10 micron bar.

2. Two mitoses in tissues of the plant *Haplopappus gracilis* which has only two chromosomes.

3. The chromosomes of the ant *Myrmecia pilosula*. The two chromosomes in a worker (A) and the single chromosome in a male (B).

107

Plants of the Same Genus Have Been Formed with 4 or 36 Chromosomes

Haplopappus gracilis is a small plant from Arizona, U.S.A. that has usually 4 chromosomes, but plants with half of this number have also been found. Within the same genus there are also species with 6, 8, 10, 12, 18, 24 and 36 chromosomes. This means that a plant with the same structural and functional features can be produced independently of whether its genes have been compressed into 4 or dispersed into 36 chromosomes (Jackson, 1973; Cremonini, 2005). When a plant is classified within the same genus it means that all the species included in it only differ, among themselves, by small morphological details.

108

A Protozoan Can Be Produced with 2 Chromosomes But Also with 500

Protozoa are living organisms consisting of a single cell. Their chromosomes have most of the features of those of higher organisms, being sometimes difficult to distinguish from human chromosomes (Fig. 1.10). The chromosomes of protozoa display an astonishing variation in chromosome number.

Spirotrichonympha polygyra has only 2 chromosomes, but many species have numbers such as: 6, 8, 12, 14, 16, 20, 30, 40, 200 until they reach one of the highest values known, 500–600 in *Amoeba proteus* (Kudo, 1971). This means that the same type of unicellular organism is produced using 2 or 500 chromosomes.

109

In Birds and Plants a Series of Minute Chromosomes Are an Obligatory Component of Their Chromosome Set

Chromosomes may become so small that they reach the limits of resolution of the light microscope. This is for instance the case in birds. They possess large chromosomes like other animal groups, but at their side they carry a plethora of minute chromosomes. These are so small that they have hardly place to have a centromere or telomeres of normal size. These minute bodies are a permanent part of the chromosome set of birds and presumably have been formed long ago since they appear in many species.

The same situation occurs in several plant families in which minute chromosomes are also a regular part of the chromosome complement coexisting at the side of large ones (Fig. 5.11).

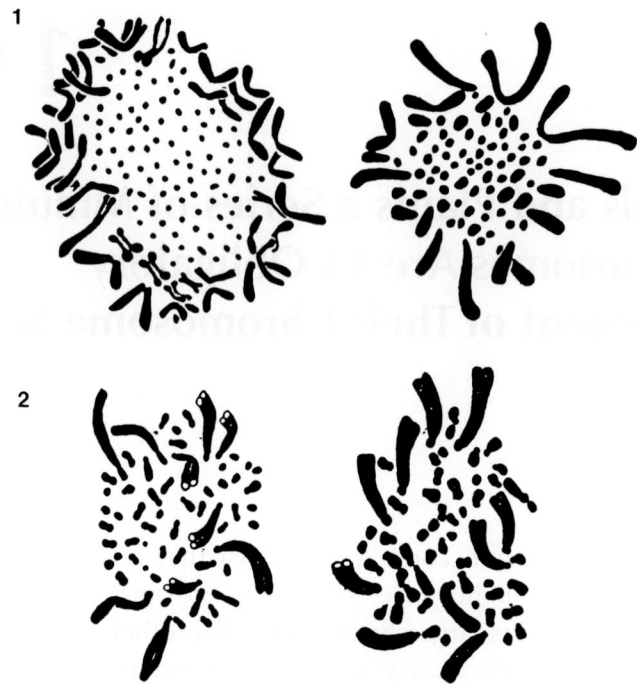

Fig. 5.11 Minute chromosomes, close to the limit of microscopical resolution, are a permanent feature of chromosome sets in animals and plants.

1. The chromosome complement of the lizard *Tupinambis teguixin* (*left*). The chromosomes of the pheasant *Syrmaticus soemmeringi* (*right*).

2. Chromosome sets of two plants: *Yucca aloifolia* (*left*) and *Agave americana* (*right*). The minute chromosomes have been preserved at the side of large ones.

110

The Separation of Chromosomes into Minor Units, as well as Their Reunion, Follows Well-Defined Solutions

Let us now look at the characteristics of this phenomenon:

(1) This chromosome ability is independent of organism complexity since it has persisted throughout evolution extending from protozoa to mammals.

(2) When the chromosome number is low their size is large. Inversely, when the chromosome number is high they tend to be small. This is especially evident in ants and deer. The single chromosome of the male of the ant *Myrmecia pilosula* has the enormous length of about 35 microns, whereas in most ant species with 18 to 36 chromosomes these measure usually 1 to 5 microns (Taber and Cokendolpher, 1988). The 3 chromosomes of *M. muntjak* are all quite large, the biggest measuring 15 microns, whereas the largest and smallest chromosomes of red deer (34 chromosomes) oscillate between 1 and 4 microns. The same size variation occurs in the 23 chromosomes of *M. reevesi*. To check if any appreciable amount of DNA had been lost in these transformations the total amount of DNA per nucleus was measured in both muntjac species giving similar values.

Thus, this process has involved little or no DNA loss (Lima-de-Faria, 1980).

(3) The battery of sophisticated molecular methods that were used to check on the possible homologies between the different chromosome constellations revealed the following: 1) Specific DNA sequences could be recognized following their transfer to different chromosomes. 2) These DNA sequences tended to occupy the same territory within the new chromosomes, i.e. they used a location, in relation to other chromosome segments, that repeated the internal relationships found within the original chromosome. For example centromeres moved to centromere locations, and proximal regions of the arms were found to occupy a corresponding location in the new chromosomes. Hence, ordered solutions dominated the transfer of DNA material (Fig. 5.9).

(4) The reassembly of minor chromosomes into large ones may have been a later event. Most placental mammals have chromosome numbers between 19 and 24 (sexual cells) (White, 1973). This means that the deer *M. reevesi* which has 23, has the predominating normal number of chromosomes. As such it is expected that the reduction to 3 chromosomes was a later event during species evolution, i.e. the large chromosomes were formed by the assembly of minor ones. This is also the case in ants where common numbers are 5 to 16 (Imai *et al.*, 1987). Thus, the single huge chromosome is considered to be a late event produced by assembly.

(5) The reverse situation also occurred. The presence of 94 chromosomes in ants, and of 500 in protozoa, is considered to be due to the disassembly of the larger units and may be the result of recent evolution.

(6) Much remains to be elucidated since many questions are still unanswered by these results. The way centromeres and telomeres behave in their new locations is a big question mark. Moreover, it is well known that the function of genes changes when their location is modified as a result of the effects of new neighboring genes.

Irrespective of what future research may disclose, the following remains established: 1) The chromosomes may divide themselves into minor units that maintain their functions. 2) Chromosomes may reunite into larger units that also maintain their functions. 3) The process is so ordered, and the action of the genes, in the novel positions remains so balanced, that practically the same organism is produced irrespective of the chromosome number that it harbors in its cells.

111

The Properties Shared by Magnets and Chromosomes May Have Their Origin in the Polarization Already Present at the DNA Level

In several respects the behavior of the chromosome resembles that of magnets. The question then arises. What mechanism lurks behind these features? The explanation seems to lie closer than one would expect, because when the chromosome is analyzed at the molecular level it turns out to be already a prisoner of an intrinsic polarity.

(1) DNA itself is a polar molecule. In the Watson and Crick model the two chains of the molecule run in opposite directions in terms of the 3'–5' phosphate-deoxyribose linkages. Thus, the two ends of each polynucleotide chain are chemically different. The result is that the double helix consists of two polynucleotides running in different directions and having opposite polarity (Fig. 5.12).

(2) This type of polarity is unique to DNA. Of the four macromolecules: DNA, RNA, protein and polysaccharides (sugars) only DNA has a double-stranded structure in which polarity is opposite in both strands. RNA has also a polarity in its nucleotides but, in contrast to DNA, most RNA molecules are single stranded forming double chains for only short

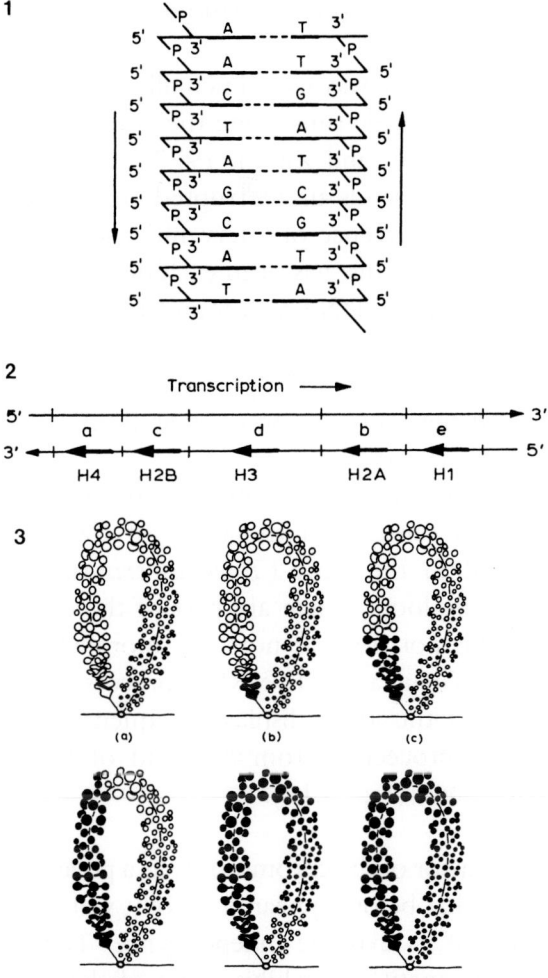

Fig. 5.12   Polar properties of DNA and of the chromosome.

1. Diagram of the DNA double helix, showing the specific pairing of the bases (A = adenine, G = guanine, T = thymine, C = cytosine). The reversed direction of the 3′–5′ phospho-diester linkages of the two chains is indicated by the descending and ascending vertical arrows (P = phosphorus. Interrupted lines between bases = hydrogen bonds).

2. Sequential location of the five histone genes (H4, H2B, H3, H2A and H1) in the chromosomes of the sea urchin *P. miliaris*. Their arrangement as well as their transcription are polarized.

3. Diagram to illustrate the progressive incorporation of tritiated-uridine in RNA on the giant granular loop of chromosome 12 of the newt *T.c. cristatus*. The radioactive RNA labelled granules are shown black. (a) Prior to injection, (b), (c), (d), (e) and (f) 1, 2, 4, 7 and 14 days after injection respectively. This means that the synthesis of RNA on the chromosome is polarized.

regions. Proteins and polysaccharides do not display this type of polarity.

(3) Already in 1973 it was experimentally demonstrated that a single DNA molecule runs along the entire length of a chromosome (Kavenoff and Zimm, 1973). This means that the DNA polarity extends uninterrupted from one end to the other since the presence of intervening non-polarized molecules, such as proteins, has been excluded.

(4) The polarity of the DNA was found to occur as well in its genes. The five histone genes of sea urchins show polarity. The coding sequences of this protein family all have the same polarity and the transcription of their RNA is also polarized (Gross *et al.*, 1976) (Fig. 5.12).

(5) Demonstration that polarity extends throughout the chromosome axis came from the work of Callan (1963). He noticed that in the newt (*Triturus cristatus*) there was a polarity in the loops which are part of the chromosomes of oocytes. This polarity was not only morphological but also functional. When radioactive RNA precursors were injected into the newt the loops displayed sequential labelling. The radioactivity proceeded from one end of the loop to the other with time ((Fig. 5.12).

When the behavior of the chromosome as a polarized unit is further elucidated, what before appeared as folly, will become evident as a vindication of its extreme independence. The chromosome does not need magnetite to exhibit some of the properties of a magnet.

References

Part V

Oersted HC. (1820) In: D Millar, I Millar, J Millar, M Millar (eds) (2002) *The Cambridge Dictionary of Scientists*, pp. 274. Cambridge University Press, Cambridge, UK.

Blakemore R. (1975). Magnetotactic bacteria. *Science* **190**: 377–379.

Frankel RB, *et al.* (1979) Magnetite in freshwater magnetotactic bacteria. *Science* **203**: 1355–1356.

Lins U, *et al.* (2006) Crystal habits and magnetic microstructures of magnetosomes in coccoid magnetotactic bacteria. *Anais Acad Bras Ciencias* **78**: 463–474.

Gould JL, *et al.* (1978) Bees have magnetic remanence. *Science* **201**: 1026–1028.

Walcott C, *et al.* (1979) Pigeons have magnets. *Science* **205**: 1027–1028.

Pitt VH. (1988). *The Penguin Dictionary of Physics*. Penguin Books, London, UK.

Loewy AG, Siekewitz P. (1970) *Cell Structure and Function*. Holt, Rinehart and Winston, London, UK.

Romer AS., Parsons TS. (1978). *The Vertebrate Body*. WB Saunders, Philadelphia, USA.

Schneider D. (1999). Some levity in physics. *Am Sci* **87**: 122–123.

Nuccitelli R, Erickson CA. (1983) Embryonic cell motility can be guided by physiological electric fields. *Exp Cell Res* **147**: 195–201.

Bloss FD. (1971) *Crystallography and Crystal Chemistry. An Introduction*. Holt, Rinehart and Winston, NY, USA.

Fukui A, Asashima M. (1994). Control of cell differentiation and morphogenesis in amphibian development. *Int J Dev Biol* **38**: 257–266.

Huxley J, de Beer GR. (1934) *The Elements of Experimental Embryology.* Cambridge University Press, Cambridge, UK.

de Robertis EA, *et al.* (1991) Gradient fields and homeobox genes. *Development* **112**: 669–678.

Fehilly CB, *et al.* (1984) Interspecific chimaerism between sheep and goat. *Nature* **307**: 634–636.

Mintz B, Illmensee K. (1975) Normal genetically mosaic mice produced from malignant teratocarcinoma cells. *Proc Natl Acad Sci* **72**: 3585–3589.

Mintz B. (1978) Mutagenized teratocarcinoma cells as probes of mammalian differentiation. Plenary Sessions Symposia, Abstracts, page 65. *XIV Int. Congress of Genetics*, Moscow.

Niebuhr E, Skovby F. (1977). Cytogenetic studies in seven individuals with an i(Xq) karyotype. *Hereditas* **86**: 121–128.

Merry DE, *et al.* (1985) Anti-kinetochore antibodies: use as probes for inactive centromeres. *Am J Hum Genet* **37**: 425–430.

Müller F, *et al.* (1991) New telomere formation after developmentally regulated chromosomal breakage during the process of chromatin diminution in *Ascaris lumbricoides. Cell* **67**: 815–822.

White MJD. (1973) *Animal Cytology and Evolution.* Cambridge University Press, UK.

Lima-de-Faria A, *et al.* (1986) DNA cloning and hybridization in deer species supporting the chromosome field theory. *BioSystems* **19**: 185–212.

Brinkley MM, *et al.* (1984) Compound kinetochores of the Indian muntjac. *Chromosoma* **91**: 1–11.

Goldoni D, *et al.* (1984) Cytogenetic studies on *Cervus elaphus* L. Constitutive heterochromatin and nucleolus organizer regions. *Caryologia* **37**: 439–443.

Scherthan H, *et al.* (1994) Comparative chromosome painting discloses homologous segments in distantly related mammals. *Nat Genet* **6**: 342–347.

Taber SW, Cokendolpher JC. (1988) Karyotypes of a dozen ant species from the SouthWestern USA (Hymenoptera: Formicidae). *Caryologia* **41**: 93–102.

Crosland WJ, Crozier RH. (1986) *Myrmecia pilosula*, an ant with only one pair of chromosomes. *Science* **231**: 1278.

Imai HT, *et al.* (1990) Notes on the remarkable karyology of the primitive ant *Nothomyrmecia macrops*, and of the related genus *Myrmecia* (Hymenoptera: Formicidae). *Psyche* **97** (3–4): 133–140.

Jackson RC. (1973) Chromosomal evolution in *Haplopappus gracilis*: A centromeric transposition race. *Evolution* **27**: 243–256.

Cremonini R. (2005) Low chromosome number angiosperms. *Caryologia* **58**: 403–409.

Kudo RR. (1971) *Protozoology*. Charles C Thomas, Illinois, USA.

Lima-de-Faria A. (1980) How to produce a human with 3 chromosomes and 1000 primary genes. *Hereditas* **93**: 47–73.

White MJD. (1973) *Animal Cytology and Evolution*. Cambridge University Press, Cambridge, UK.

Imai HT, *et al.* (1987) Chromosomal polymorphism in the ant *Myrmecia* (*pilosula*) n = 1. Annual Report No 38 National Institute of Genetics, Mishima, Japan, 1988.

Kavenoff R, Zimm BH. (1973) Chromosome-sized DNA molecules from *Drosophila*. *Chromosoma* **41**: 1–27.

Gross K, *et al.* (1976) Molecular analysis of the histone gene cluster of *Psammechinus miliaris*: III. Polarity and asymmetry of the histone-coding sequences. *Cell* **8**: 479–484.

Callan HG. (1963) The nature of lampbrush chromosomes. In: GH Bourne & JF Danielli (eds), *International Review of Cytology*, pp. 1–34. Academic Press, NY, London.

Sources of Illustrations

Part V

5.1 (1) and (2) Boutaric A. (1938) *Precis de Physique*. G Doin Editeurs, Paris, France (Figs. on pages 864 and 867).

(3) Hey T, Walters P. (2003) *The New Quantum Universe*. Cambridge University Press, Cambridge, UK (Fig. 12.20, page 271).

5.2 (1) *Nationalencyclopedin*. (1991) Förlaget Bra Böcker, Höganäs, Sweden (Fig. on page 384).

(2), (3) and (4) Lima-de-Faria A. (1995) *Biological Periodicity. Its Molecular Mechanism and Evolutionary Implications*. JAI Press, Greenwich, Connecticut, USA (Fig. 5, page 132).

5.3 (1) Frankel RB, *et al.* (1979) Magnetite in freshwater magnetotactic bacteria. *Science* **203**: 1355–1356 (Fig. 1, page 1356, c. 1979, AAAS).

(2) Lins U, *et al.* (2006) Crystal habits and magnetic microstructures of magnetosomes in coccoid magnetotactic bacteria. *Anais Acad Bras Ciencias* **78(3)**: 463–474 (Fig. 1, page 466).

5.4 Nuccitelli R, Erickson CA. (1983) Embryonic cell motility can be guided by physiological electric fields. *Exp Cell Res* **147**: 195–201 (Fig. 3, page 198).

5.5 Gilbert SF. (2000) *Developmental Biology*. Sinauer Associates, Inc Publishers, Sunderland, MA, USA (Fig. 3.15, page 59).

5.6   Fehilly CB, *et al.* (1984) Interspecific chimaerism between sheep and goat. *Nature* **307**: 634–636 (Figure on cover of this issue of Nature).

5.7   Gilbert SF. (2000) *Developmental Biology.* Sinauer Associates, Inc Publishers, Sunderland, MA, USA (Fig. 4.19, page 98).

5.8   (1)   Whiteley D, Nixon M. (1972) *The Oxford Book of Vertebrates.* Oxford University Press, Oxford, UK (Figs. 1 and 2, page 149).

        (2)   Wurster DH, Benirschke K. (1970) Indian Muntjac, *Muntiacus muntjak*: A deer with a low diploid chromosome number. *Science* **168**: 1364–1366 (Fig. 1, page 1365).

        (3)   Hsu TC, Benirschke K. (1971) *An Atlas of Mammalian Chromosomes. Vol 2 Folio 88.* Springer Verlag, Heidelberg, Germany (Fig. in Folio 88).

5.9   Lima-de-Faria A, *et al.* (1986) DNA cloning and hybridization in deer species supporting the chromosome field theory. *BioSystems* **19**: 185–212 (Fig. 18, page 210).

5.10   (1)   Michailova P. (1976) Cytotaxonomical diagnostics of species from the genus *Cricotopus* (Chironomidae, Diptera). *Caryologia* **29**(3): 291–306 (Fig. 1, page 294).

        (2)   Lima-de-Faria A, Jaworska H. (1964) Haplo-diploid chimaeras in *Haplopappus gracilis. Hereditas* **52**: 119–122 (Fig. 3 and 4, page 119).

        (3)   Crosland MWJ, Crozier RH. (1986) *Myrmecia pilosula*, an ant with only one pair of chromosomes. *Science* **231**: 1278 (Fig. 1, page 1278).

5.11   (1)   White MJD. (1973) *Animal Cytology and Evolution.* Cambridge University Press, Cambridge, UK (Fig. 2.3, page 67; Fig. 15.1, page 552).

        (2)   Darlington CD. (1956) *Chromosome Botany.* George Allen and Unwin, London, UK (Fig. 31, page 98).

5.12   (1)   Hayes W. (1968) *The Genetics of Bacteria and Their Viruses.* Blackwell, Oxford. (Fig. 58, page 236).

        (2)   Gross K, *et al.* (1976) Molecular analysis of the histone gene cluster of *Psammechinus miliaris*: III polarity and asymmetry

of the histone-coding sequences. *Cell* **8**: 479–484 (Fig. 4, page 482).

(3)   Callan HG. (1963) The nature of lampbrush chromosomes. In: GH Bourne & JF Danielli (eds). *International Review of Cytology*, vol 15, pp. 1–34. Academic Press, NY. (Fig. 10, page 23).

Biological Order is the Product of Self-Assembly and Self-Assembly is the Product of Atomic Recognition

112

How Clocks and Other Machines Differ from Cells

A machine is assembled by adding components which are not structurally related. So long as one achieves the desired functional result any pieces of material may be put together irrespective of how disparate the individual components may be. Hence, a rubber tube is put together with a metal screw which in turn may be connected with a plastic wheel. The plastic, the rubber and the iron pieces are not functionally adapted to each other. They have not gone through a process of interactions and recognitions. A machine needs only to have a coherent organization at the level of its final function.

The difference between a machine and a cell is particularly evident when one takes them both apart.

Anything which is man made, is easy to distinguish from a system created by natural processes. A machine taken to pieces becomes a chaotic collection of parts. If these in turn are further divided, the original pattern cannot be recognized. An organism when broken into pieces shows an order of organization at every level. Every piece, however small it may be, shows an organized texture. Moreover, every cellular molecule or structure is an integral part of the whole. The painstaking and successive building of the cell from inside out, atom by atom, molecule by molecule, did not permit anything else but a rigid fitting of every structure into the whole. A similar point of view has been formulated by

Roman Vishniac (cited by Feininger, 1956). 'Everything made by human hands looks terrible under magnification — crude, rough, and asymmetrical. But in nature every bit of life is lovely. And the more magnification we use, the more details are brought out, perfectly formed, like endless sets of boxes'.

The proof of the difference between machines and organisms resides in the fact that cell organelles, such as ribosomes and chromosomes, and organisms, such as hydras and slime molds, can self-assemble, but the components of a machine cannot. The nucleic acid and the proteins of a ribosome once isolated can recognize each other and produce by themselves a complete and functional ribosome. The separated cells of a slime mold can also find each other and build a complete organism. However, if one were to isolate the components of a clock, or of a motor and mix them subsequently, none of them would be able to recognize the other parts in the right configuration and sequence to enable them to produce alone a functional clock or motor. Every component is a foreigner, having no knowledge of the whole. This is why a machine neither reproduces nor dies.

The understanding of how cells differ from machines helps us in finding out how cells and chromosomes originated.

113

Definition of Self-Assembly and Its Basic Properties

Self-assembly is probably the most basic mechanism in living functions, but it is only recently that its significance has started to be realized. The number of cellular processes found to take place by this event is increasing all the time. Self-assembly is a term created by biochemists to describe "the spontaneous aggregation of multimeric biological structures involving formation of weak chemical bonds between surfaces with complementary shapes" (King and Stansfield, 1997).

The information resulting in the formation of complex assemblies of macromolecules "must be contained in the subunits themselves, since under appropriate conditions, the isolated subunits can spontaneously assemble in a test tube into the final structure" (Alberts *et al.*, 1994).

Polypeptide chains possess the intrinsic property of being able to change the one-dimensional information inherent in their amino acid sequence into three-dimensional configurations by self-assembly. As Lehninger (1975) points out, this change of structure "is not imposed on it by external forces: it is the inevitable and automatic outcome of the tendency of the surrounding water molecules to seek the state of maximum entropy and the tendency of the polypeptide chain to seek its state of minimum free energy". Hence, self-assembly is inevitable and automatic due to the chemical properties of water, which is the main component of all

Level	Example	Reference
Elementary particles	Mesons result from union of pairs of quarks and antiquarks.	Pagels, 1982
Atoms	Protons, neutrons, and electrons self-assemble into atoms.	Pitt, 1988
Crystals	Atoms self-assemble into crystals by ionic, covalent, metallic and molecular bonding.	Jaffe, 1988
Proteins	Aspartate transcarbamoylase; the 12 units, which have been dissociated, are able to reconstitute the enzyme.	Bothwell and Schachman, 1974
Membranes	Nuclear membranes self-assemble from cell-free extracts of eggs.	Dabauvalle et al., 1991
Ribosomes	The 3 RNAs and 53 proteins of ribosomes have been isolated and reassemble into active particles. The information for correct self-assembly is present in the structure of the molecules.	Spirin, 1986
Nucleosomes	Purified DNA and histones self-assemble into the essential component of chromosomes, called nucleosome.	Oudet et al., 1975
Viruses	The isolated RNA and proteins of tobacco mosaic virus self-assemble, thus producing an infectious virus.	Mindich et al., 1982
Organisms	Self-assembly of the dispersed amoebae of *Dictyostelium discoideum* results in a complete slime mold.	Bonner, 1983
Organisms	Hydras can self-assemble from their dispersed cells.	Gierer, 1974
Organisms	After dissociation, marine sponges produce a whole organism from their cells.	Müller et al., 1976
Organs of Mammals	Dispersed liver cells, kept in tissue culture, self-assemble. The resulting liver reacquires its function.	Moscona, 1959
Organs of Mammals	Dissociated cells of rat gonads reorganize into testes and ovaries.	Ohno et al., 1978
Human tissues	Human skin has been produced by allowing dispersed cells to self-assemble.	Dubertret et al., 1987
Human Organs	Human capillary endothelial cells from tissue cultures are able to self-assemble into blood capillaries.	Folkman and Haudenschild, 1980

Fig. 6.1   Self-assembly extends from elementary particles to living organisms.

"The visible world is the result of the organization of energy" as the American physicist Pagels (1982) put it. At the bottom of the universe's construction is the combination capacity of sub-elementary particles, such as the quarks, which associate into mesons and other particle families. This self-assembly capacity extends to the building of atoms, crystals and macromolecules. Since it is a property inherent to matter it does not stop here, but is equally evident in the formation of organisms and human organs. It simply could not be altered during evolution.

Note: The self-assembly of isolated RNA and protein in tobacco mosaic virus was obtained by Fraenkel-Conrat (1962) as mentioned in the text and depicted on Fig. 6.3. The work of Mindich *et al.* (1982) is a confirmation of the same phenomenon in viruses whose genetic material is DNA.

The references in this table are found in Lima-De-Faria 1995 (see sources of illustrations).

cells, and to the chemical properties intrinsic to the macromolecule in question.

Physicists have long dealt with the spontaneous and intrinsic ability of elementary particles to assemble in the formation of other particles and atoms. To them the organization of protons, neutrons and electrons into atoms was long ago considered to be a natural process.

One way in which atoms show their capacity for self-assembly is in the spontaneous formation of crystals. Most mineral structures build large crystals that can be easily seen with the naked eye. Under other conditions the minerals that surround us seem at first sight to be disordered but when analyzed with the microscope their pattern becomes evident. The crystals are the messengers of the atomic order which is hidden in matter, since they display it in an unequivocal way.

A supramolecular structure consists of subunits which in turn are formed by other minor subunits. During the formation of macromolecules some molecules have a more important role than others in the sequence of the assembly. Ribosomes, viruses and other supramolecular structures, which are formed by recognition processes, often show a hierarchical organization.

The complicated structures that arise in the cell, with specific biological functions, are the product of information which is inherent in the molecules themselves (Becker *et al.*, 2003) (Fig. 6.1).

Self-assembly turns out to be: spontaneous, inherent, inevitable, automatic and hierarchic. In the following pages these five main properties will be described in detail.

114

The Mechanism Responsible for Self-Assembly is Independent of External Information

The most significant aspect of self-assembly from the point of view of molecular and chromosomal evolution is that the ordered aggregation of molecules, macromolecules and supramolecular systems takes place independently of any external information. Not only enzymes, ribosomes and viruses can self-assemble due to the information inherent to their structures but cell membranes also possess this capacity. Phospholipid molecules contain within themselves the necessary information to assemble into bilayer systems and they perform this process in the absence of other sources of information. The amino acid sequences which build polypeptide chains also carry inherent information.

This is exemplified by hemoglobin, which has been studied intensively. Normal adult hemoglobin is formed of two alpha and two beta polypeptide chains. When alpha and beta chains are mixed in solution one would expect, on a random basis, that the two types of subunits would form aggregates of random numbers of subunits. This is not what happens. Under normal conditions hemoglobin molecules do not bind to each other to form four alpha, four beta, one alpha and three beta or three alpha and one beta geometric systems. Hemoglobin always appears in the form of two alpha and two beta subunits (Fig. 6.2).

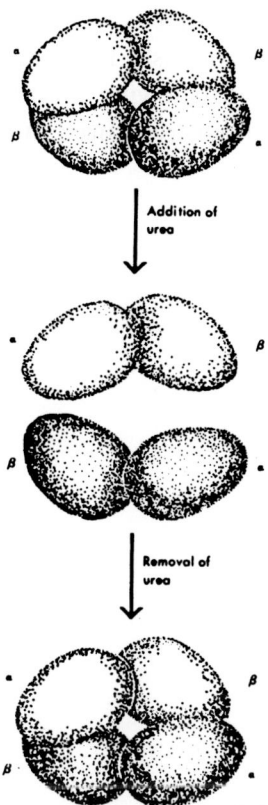

Fig. 6.2   Self-assembly of macromolecules.

Formation of an active hemoglobin molecule from two half-molecules. Each hemoglobin molecule contains two α and two β chains. When placed in urea (a reagent which destabilizes weak bonds), the native molecule falls into two halves, containing one α and one β chain. Upon removal of the urea, the halves reassociate to form the complete molecule.

Another demonstration of the impressive specificity of the inherent information is furnished by an experiment in which native hemoglobins from different mammalian species were mixed under favorable chemical conditions. They did not exchange subunits in any appreciable amount to form hybrid configurations (Lehninger, 1975). Therefore, randomness and external information are processes which seem to be foreign to the mechanisms which shape the supramolecular systems.

As Rose *et al.* (2006) stated: "The conclusion that proteins can self-assemble spontaneously is based on Anfinsen's Nobel prize-winning experiments showing that the protein ribonuclease can be reversibly denatured/renatured in a test tube" (Anfinsen, 1973).

115

The Self-Assembly of an Enzyme is So Rapid that It Takes Less Time than the Synthesis of Its Component Polypeptides

Another characteristic of self-assembly is its impressive speed. Bothwell and Schachman (1974) have studied in detail the enzyme aspartate transcarbamoylase, which is composed of 12 polypeptide chains, of which six are catalytic and six are regulatory. Once again randomness is excluded. The first step in assembly is the association of one catalytic with one regulatory chain. Due to the specificity and the regularity of the aggregation in the binding of the 12 subunits, the process of assembly is extremely rapid. The entire enzyme is built up from the catalytic and regulatory subunits in less time that it takes to synthesize the original polypeptide chains.

Enzymes can function very quickly due to the fact that they attach to their substrates by weak rather than strong bonds. This allows an association to be made rapidly and just as quickly broken apart. The speed attained may be very high; an enzyme may function as many times as 10^6 times per minute (Watson, 1976).

116

No One Believed in the Self-Assembling Capacity of Viruses

When Fraenkel-Conrat contemplated the idea of self-assembling virus particles from their constituents — the nucleic acids and the proteins — no one would believe him. How could such a complex structure as an RNA virus self-assemble? The answer came quickly; the tobacco mosaic virus, with a mass of about 40,000 kdal. and dimensions of $160 \times 3,000$ Å, self-assembled without difficulties. These are dimensions similar to those of the smallest chromosomes found regularly in bird genomes. When the isolated RNA is added to the isolated protein, under appropriate pH conditions, they start to assemble and the virus particles formed are fully infectious and indistinguishable from the original virus (Fraenkel-Conrat, 1962). The self-assembly occurs spontaneously in solution; no additional molecular component is necessary (Fig. 6.3).

The process has been further studied, its specificity and speed being described by Butler (1999) and Kegel and van der Schoot (2006).

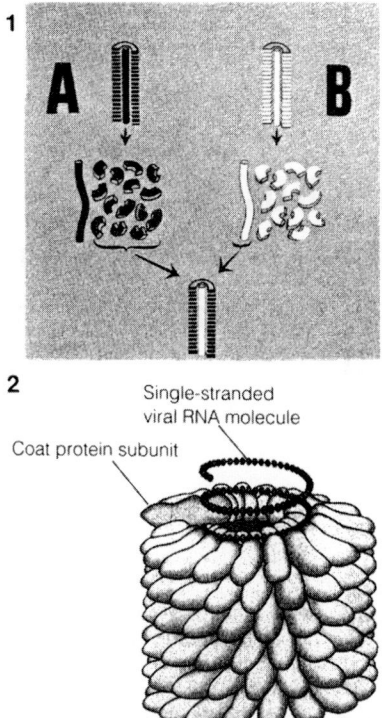

Fig. 6.3   How to produce a virus from its molecular constituents.

A and B represent two strains of tobacco mosaic virus. Both were degraded into protein and RNA (represented as circular segments and a pipelike body, respectively). Reconstitution of the protein from one strain with RNA from the other strain yields a mixed virus. This, upon inoculation, yields disease symptoms and its progeny isolated from tobacco plants is indistinguishable from the parent virus. Below, a diagram of the packing of the viral RNA with the protein subunits.

117

The Bacterial Virus *T4* Has a Programmed Pathway of Assembly that Has Been Described in the Utmost Detail

More than half of the genes present in the T4 DNA contribute to the formation of the complete virus particle. Each gene codes for a specific protein subunit. The other viral genes produce the enzymes required for viral DNA synthesis and host cell-wall lysis. The viral genes contain the code for the protein subunits but self-assembly is responsible for putting these proteins together and around the DNA, building the large molecular edifice called a virus particle.

Self-assembly in itself is an ordered process since it involves molecular specificity and perfect recognition, but in T4 there are other types of order involved in its assembly. There is a specific sequence in the assembly process and there are three separate assembly lines during the formation of the infectious virus particles. Intact heads, tails and fibers are formed separately and only later are they united to form the complete virus. This was shown by Wood and Edgar (1967), who reconstituted *in vitro* the bacteriophage T4 by mixing mutant viral particles lacking tails with an extract from cells infected with a virus defective in head proteins (Fig. 6.4). Sixteen genes participate in the formation of the heads and 15 in

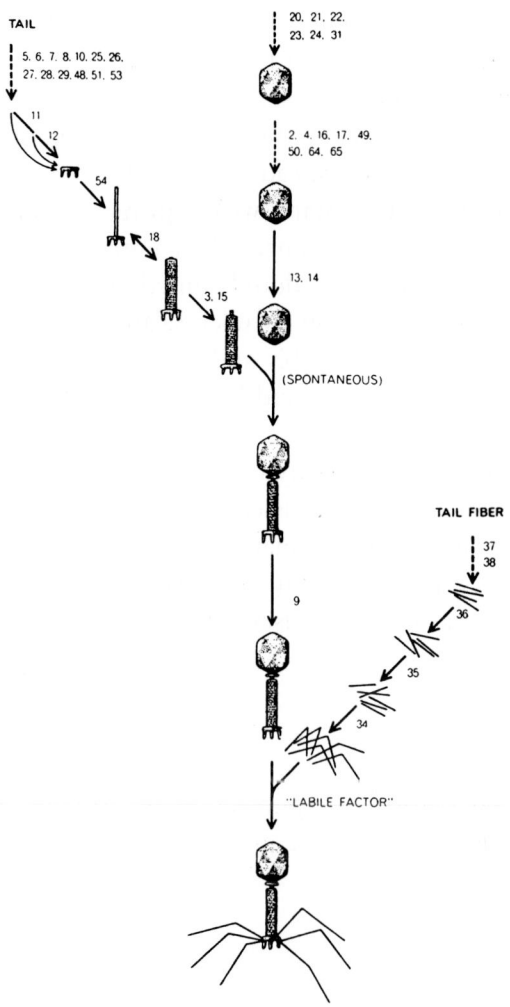

Fig. 6.4   Reconstitution *in vitro* of the bacteriophage T4 (a virus which infects bacteria).

The morphogenetic pathway has three principal branches leading independently to the formation of heads, tails and tail fibres, which then combine to form complete virus particles seen at the bottom of the figure. The numbers refer to the gene product or products involved at each step. The solid portions of the arrows indicate the steps that have been shown to occur in extracts. The capacity for self-assembly could not be more evident.

the organization of the tails. The heads and tails then combine spontaneously. There is also order in the assembly of the tail itself. The tail cannot begin to be assembled from its middle part. There is an obligatory sequence in which each step promotes the occurrence of the next step in a cascade process. It is now known that in T4 the prohead assembly starts with a primer or initiator protein which governs the 5-fold symmetry. The initiator is supposed to be the gp20 protein which has a double function. It is also required to pack the DNA into the head (Müller-Salamin *et al.*, 1977; Hsiao and Black, 1977). The assembly and disassembly of T4 have been further studied by Caldentey and Kellenberger (1986) and Arisaka (2001).

The self-assembling capacity of viruses is now being used in an effort to direct the ordered assembly of foreign molecules. The DNA virus M13 can tag and manipulate inorganic molecules such as metals. This capacity is being employed to guide the formation of semiconductors and other useful products (Ross, 2006).

118

The Self-Assembly of Ribosomes
Has Been Obtained in the Test Tube

Ribosomes are the sites of protein synthesis.

Of all cell organelles, ribosomes are those in which self-assembly has been studied most thoroughly. The 70S ribosome particles of the bacterium *E. coli* consist of two subunits, 30S and 50S (Fig. 6.5). The 30S subunit contains one molecule of 16S RNA and 21 different proteins. The 50S subunit has 5S RNA and 23S RNA associated with 34 different proteins. The three RNAs have been isolated and most proteins purified. Functionally active ribosomes of bacteria could be self-assembled from a mixture of the purified molecular components. All the information for correct self-assembly is present in the structure of the component molecules. It was found that the order of assembly of the proteins follows a specific sequence and that the addition of one protein enhances the incorporation of the next (Nomura, 1973; Röhl and Nierhaus, 1982; McGrath and Butler, 1997; Liljas, 2004).

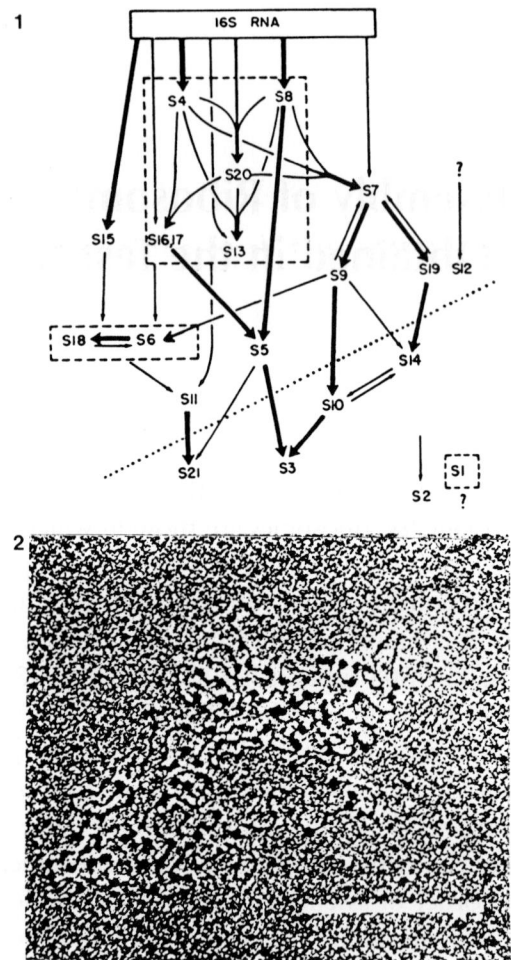

Fig. 6.5 Self-assembly of ribosomes and nucleosomes.

1. The ribosome particles of the bacterium *E. coli* consist of two subunits, 30S and 50S. The 30S subunit contains one molecule of 16S RNA and 21 different proteins. Their assembly map is shown above. Arrows between proteins indicate the facilitation effect on binding of one protein on another; a thick arrow indicates a major interaction effect.

2. Electronmicrograph of nucleosomes resulting from the self-assembly of their four histone proteins and DNA. Nucleosomes are the major components of the chromosome structure.

119

Self-Assembly of the Chromosome Fiber and of Other Chromosome Structures Involved in Its Movement

The nucleosomes, which are the main constituents of chromosomes of higher organisms, have also been self-assembled. The nucleosomal pattern of spherical particles can be reconstituted *in vitro* from purified DNA and the isolated four histone proteins that are part of their structure. This occurs irrespective of the cellular origin of the DNA or the histones (Oudet *et al.*, 1975). Laskey *et al.* (1978) have stressed that: "the spatial information required for histone-histone interactions and histone-DNA interactions resides in the histones and DNA themselves" (Fig. 6.5).

Camerini-Otero *et al.* (1977) have reconstituted DNA with essentially all possible combinations of histone types. The association of histones and DNA *per se* was not sufficient to give rise to nucleosomes. Only H3 and H4 histones were able, together with DNA, to form a nucleoprotein complex with many of the structural features of the nucleosome. This demonstrates the chemical specificity of the self-assembly reaction (Yoshikawa *et al.*, 2001).

Not only the nucleosomes, but also the centromeres, with their important function of attaching to the spindle fibers and guiding the chromosomes in their movements, have also been self-assembled in yeast cells. The assembly of the multiprotein complexes that build the active region of the centromere is hierarchical, as in the case of viruses and ribosomes (Wulf *et al.*, 2003).

Hence, the essential components of the chromosome have in themselves the physico-chemical information necessary to guide their self-assembly. This is evidence of the auto-canalization inherent to the structure of the chromosome.

120

Single Sponge Cells Have the Information to Produce a Whole Organism

The cells of marine sponges can be dissociated by pushing them through fine bolting cloth. The cells settle at the bottom of the vial but soon they start to aggregate by means of their amoeboid movements. After 4 days complete sponges with canals and flagellated chambers emerge. The chemical responsible for self-assembly has been isolated. This substance is species specific since it does not cause the self-assembly of sponges from other groups (Galtsoff, 1923; Müller *et al.*, 1976).

121

A Hydra with Its Highly Complex Tissues Can Self-Assemble from Dispersed Cells

Not only sponges but other species consist of cells which, once isolated, can self-assemble to produce the entire organism.

The phenomenon has been studied in detail in *Hydra* by Gierer (1974) (Fig. 6.6). These freshwater coelenterates are disaggregated into cells by pipetting the animals repeatedly in a solution of high salt concentration. The pieces are passed through filters until single cells are obtained, which are then allowed to self-assemble in a dish containing a suitable medium. After a day ectodermal and endodermal cells emerge, and by the second day tentacles are formed.

Hydras can reproduce both asexually and sexually and they have a polar structure characterized by the presence of a head and a foot. Each individual consists of 100,000 cells of a dozen different cell types. The animal has a nervous system concentrated mainly in the head and is provided with highly differentiated cells called nematocytes. These are stinging cells that migrate to the tentacles where they attack and kill the animal's prey. They are considered to be among the most complex cells known to biologists. They consist of a capsule, a receptor bristle, a filament and a barb. When the receptor bristle is touched, the

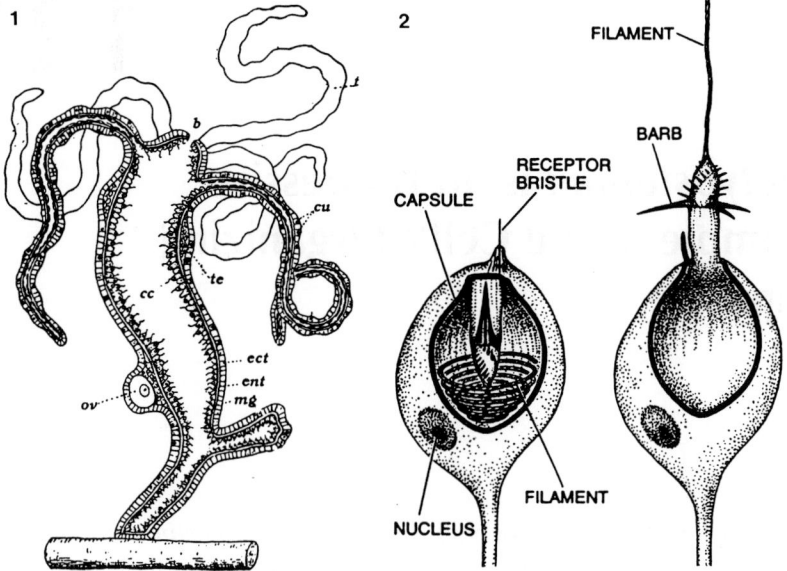

Fig. 6.6   The self-assembly of an animal from its dispersed cells.

1. The organization of hydra in a schematic cross-section: cc, digestive cavity; cu, nemato-cytes; ect, ectoderm; ent, endoderm; b, hypostome; ov, ovule; t, tentacle; te, testis. Such a complex organism can be disaggregated into its constituent cells. Within days the single cells self-assemble producing the original hydra.

2. In hydras there are several types of nematocytes which have a highly complex structure and function. Before these stinging cells are discharged they contain a long coiled filament. When the receptor bristle is touched, an impulse is relayed to a capsule, which expels the filament (right). The filament releases a stinging substance, and the barb at its base helps to impale the prey, which is eventually carried to the animal's mouth.

capsule expels the filament which releases a stinging substance. The barb helps to capture the prey. Hence, an animal with a nervous system and such complex organs can be formed in its entirety by self-assembly of its isolated cells (Fig. 6.6).

122

Dictyostelium is an Impressive Example of the Cell's Organizing Capacity

The life cycle of the amoeboid slime mold *Dictyostelium discoideum*, is very short, it lasts for only 4 days (Fig. 6.7). Each individual possesses a fruiting body which, when mature, releases many spores. Under suitable conditions each of these spores opens up and hatches one unicellular amoeba. During the vegetative state the amoeboid cells are separate and free swimming. When the amoebae become sufficiently numerous and the food supply is depleted the aggregation of the cells starts. The amoebae stream radially toward central collecting points, and as they reach the center, a conical mound of cells is built up (Bonner, 1947; Gerisch, 1968). The cell mass elongates and contains between a few hundred to 100,000 cells. A molding and sculpturing of the cellular mass follows the aggregation process, and the wormshaped slug migrates by gliding at a speed of 2 mm per hour. This movement is due to the fact that the individual amoebae which constitute the organism are themselves in active motion. Finally, during the fruiting period, the whole cell mass rights itself and grows up into the air to form a delicate stalk supporting at its apex a round, smooth, spore body (Bonner, 1952). The fruiting body consists of three distinct cell groups: spores, stalk cells and basal disc cells. These have different sizes, shapes, constitutions and functions, as do the various tissue cells of, for instance, a frog embryo (Bonner, 1959).

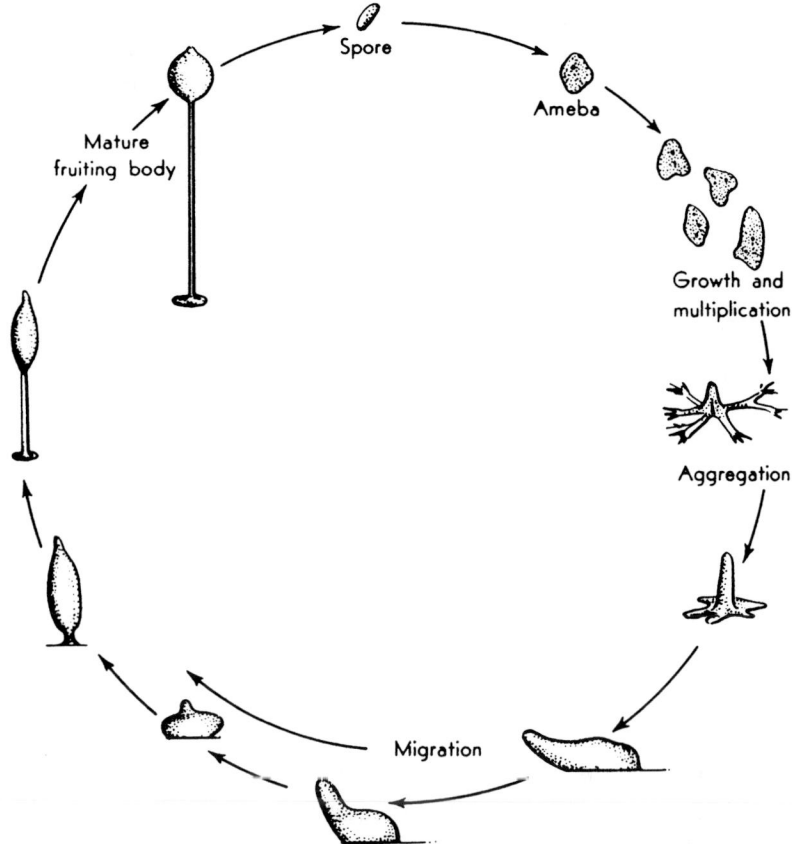

Fig. 6.7   The life cycle of the amoeboid slime mold *Dictyostelium discoideum.*

Each individual possesses a fruiting body which releases many spores (in the circle at 11 o'clock). Under suitable conditions each of these spores opens up and hatches one unicellular amoeba. During the vegetative state the amoeboid cells are free swimming. When the food supply is depleted the aggregation of the cells starts. The amoebae stream radially toward a central collecting point where hundreds of cells assemble. Then the resulting wormshaped slug migrates. Finally, during the fruiting period, the whole cell mass rights itself and forms a delicate stalk supporting at its apex a round, smooth, spore body.

A protein has been identified which is present on the surface of cells with the capacity for self-assembly but which does not occur on mutant cells unable to aggregate. This protein, named *discoidin*, is supposed to act as a specific cell-adhesion molecule that binds *D. discoideum* cells together by reacting with specific receptors on adjacent cells (Gregg and Yu, 1975). Since then another protein, called myosin, has been found to be involved in the assembly of this organism (Kuczmarski and Spudich, 1980; de la Roche *et al.*, 2002).

123

The Self-Assembly of Cells Leading to Tissue Formation is Like the Precipitation of Crystals Out of a Solution

It was only in the last decades that it became possible to dissociate tissues into cells.

This was achieved by dissolving the viscous cement that binds vertebrate cells. Such a technique led to attempts to recombine the individualized cells into tissues.

Most embryonic cells are unable to proceed with their development when isolated. They may continue to divide but lose their specific pattern. Chick embryonic kidney tissue is formed by various cell types organized into tubules. When the kidney tissue is dissociated into single cells these cannot display their characteristics outside the multicellular formation.

Once the cells are allowed to reconstruct a tissue, the individualized cells can self-assemble without any other information, rebuilding a tissue with the same organization and function. As Moscona (1959) points out, the cells condense into tissue clusters "Like crystals precipitating out of a solution".

If a similar experiment is carried out with liver cells, these reassemble into structures identical to the lobules of the intact liver and accumulate glycogen, i.e. reacquire the initial function.

Heart cells also aggregate into tissue that contracts rhythmically.

Mixed cells from different organs and from different organisms have the ability to recognize one another and to build separate organs. Mouse cartilage-forming cells, when combined with chick kidney-forming cells, group themselves in such a way that the mouse cells reconstruct cartilage and the chick cells kidney tubules.

Human capillary endothelial cells from long-term cultures were also able to form tubular networks *in vitro*. These appear almost identical under the light and electron microscope, to capillary vascular beds found *in vivo* (Folkman and Haudenschild, 1980; Miao *et al.*, 2005). Human skin has been produced by allowing dispersed cells to self-assemble (Dubertret *et al.*, 1987).

The cells, like the atoms, cannot be confused by being mixed together. Due to their specific molecular properties they reassemble only according to their construction.

124

The Unfailing Power and Accuracy Inherent to Self-Assembly — The Nuclear Envelope Has Reassembled with Precision during an Untold Number of Cell Divisions

In bacteria there is no nucleus. The chromosomes are free in the cytoplasm without any membrane separating them from the rest of the cell contents. The nuclear envelope is a later invention which imprisoned the chromosomes in the nucleus. It appeared with the emergence of protozoa and algae 600 million years before present (David *et al.*, 2002).

Where does the nucleus come from? No one seems to know. The nucleus has no evolutionary history or a very rudimentary one. For this reason there are no sweeping hypotheses on its ancestry as there are for other organelles such as the mitochondria or the chloroplasts. These are supposed to have originated from symbiotic bacteria, an interpretation which is supported by molecular evidence (Dickerson, 1980).

For the nucleus the situation is totally different. The nuclear envelope of most cells of higher organisms is disrupted and disintegrated at every cell division. Analysis in the electron microscope discloses that the nuclear envelope is a double membrane that is

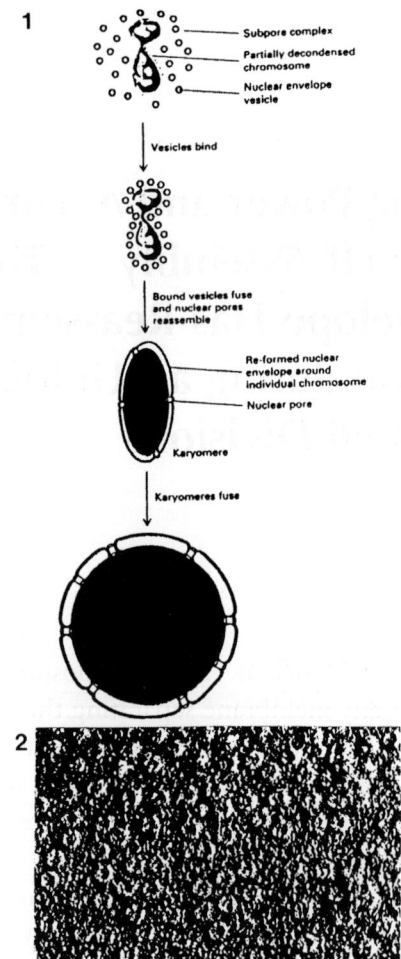

Fig. 6.8 Assembly of the nuclear envelope at the end of cell division.

The cell nucleus is limited by a membrane that encircles the chromosomes. Well-defined pores cover the surface of the nuclear envelope. The envelope disassembles at every cell division, liberating the chromosomes into the cytoplasm and later self-assembles (at the end of cell division) imprisoning them again within its boundaries.

1. Stages of self-assembly. Subpore complexes and nuclear envelope vesicles become partially condensed around a chromosome. Subsequently bound vesicles fuse and nuclear pores reassemble into groups that finally merge rebuilding the envelope.

2. Electronmicrograph of the nuclear envelope of the fly *Drosophila* salivary gland cells showing the regular structure and distribution of the nuclear pores.

continuous with the endoplasmic reticulum and is perforated by innumerous pores with a well defined structure and size.

At every cell division, when the nuclear envelope breaks down, the chromosomes are released from the nucleus into the cytoplasm becoming free to move in the new medium. After they reach the spindle poles, and come close to each other, they build two separate chromosome sets at the cell's opposing ends. Then, they are enclosed again by the nuclear envelope that is formed anew (Vigers and Lohka, 1991) (Fig. 6.8).

The disassembly and self-assembly of the nuclear envelope is: 1) Timed in relation to the other cell events. 2) Depolymerization as well as self-assembly are controlled by the protein *lamin*. This in turn is dependent, for its action, on two cyclins and other proteins, that direct most of the sequence of stages occurring during cell division (Minshull *et al.*, 1989). These events take place at every cell division in plants and animals and the envelope has the same structure in algae and humans.

The number of cell divisions that have taken place in the building of the myriads of organisms that have emerged in the last 600 million years, is such a phenomenal number, that it may be compared to the number of stars in the Universe. At every one of these cell divisions the envelope's components reassembled in a way that produced a new membrane without the intervention of external factors. They only used the molecular edifice of the cell.

It may be difficult to find a better example of the independence, permanence, accuracy, and power of self-assembly.

References

Part VI

Pagels HR. (1982) *The Cosmic Code*. Michael Joseph, London, UK (cited on Fig. 6.1).

Feininger A. (1956) *The Anatomy of Nature*. Crown Publ Inc, NY, USA.

King RC, Stansfield WD. (1997) *A Dictionary of Genetics*. Oxford University Press, Oxford, UK.

Alberts B, *et al*. (1994). *Molecular Biology of the Cell*. Garland Publ Inc, NY, USA.

Lehninger AL. (1975) *Biochemistry*. Worth Publ Inc, NY, USA.

Becker WM, *et al*. (2003) *The World of the Cell*. Benjamin Cummings, San Francisco, USA.

Rose GD, *et al*. (2006) A backbone-based theory of protein folding. *Proc Nat Acad Sci* **103**(45): 16623–16633.

Anfinsen CB. (1973) Principles that govern the folding of protein chains. *Science* **181**: 223–230.

Bothwell M, Schachman HK. (1974) Pathways of assembly of aspartate transcarbamoylase from catalytic and regulatory subunits. *Proc Nat Acad Sci* **71**: 3221–3225.

Watson JD. (1976) *Molecular Biology of the Gene*. WA Benjamin, Menlo Park, CA, USA.

Fraenkel-Conrat H. (1962) *Design and Function at the Threshold of Life: The Viruses*. Academic Press, NY, USA.

Butler PJG. (1999) Self-assembly of tobacco mosaic virus: the role of an intermediate aggregate in generating both specificity and speed. *Ph Trans Royal Soc London B, Biol Sci* **354**(1383): 537–550.

Kegel WK, van der Schoot P. (2006) Physical regulation of the self-assembly of tobacco mosaic virus coat protein. *Biophysical Journal* **91**(4): 1501–1512.

Wood WB, Edgar RS. (1967) Building a bacterial virus. *Sci Am* **217**(1): 60–74.

Müller-Salamin L, *et al.* (1977) Localization of minor protein components of the head of bacteriophage T4. *J Virol* **24**(1): 121–134.

Hsiao CL, Black LW. (1977). DNA packaging and the pathway of bacteriophage T4 head assembly. *Proc Nat Acad Sci* **74**(9): 3652–3656.

Caldentey J, Kellenberger E. (1986) Assembly and disassembly of bacteriophage T4 polyheads. *J Mol Biol* **188**(1): 39–48.

Arisaka F. (2001) Self-assembly and phage-encoded molecular chaperones of bacteriophage T4. *Virus (Nagoya)* **51**(1): 57–62.

Ross PE. (2006) Viral nano electronics. *Sci Am* October 2006: 31–33.

Nomura M. (1973) Assembly of bacterial ribosomes. *Science* **179**: 864–873.

Röhl R, Nierhaus KN. (1982). Assembly map of the large subunit (50S) of *Escherichia coli* ribosomes. *Proc Nat Acad Sci* **79**: 729–733.

McGrath KP, Butler MM. (1997) Self-assembling protein systems: a model for materials science. *Bioeng Mat* **Protein-based Materials**: 251–279.

Liljas A. (2004) *Structural Aspects of Protein Synthesis*. World Scientific Co, USA.

Oudet P, *et al.* (1975) Electron microscopic and biochemical evidence that chromatin structure is a repeating unit. *Cell* **4**: 281–300.

Laskey RA, *et al.* (1978) Nucleosomes are assembled by an acidic protein which binds histones and transfers them to DNA. *Nature* **275**: 416–420.

Camerini-Otero RD, *et al.* (1977) The structure of the nucleosome: evidence for an arginine-rich histone kernel. In: HJ Vogel (ed), *Nucleic Acid Protein Recognition*, pp. 151–158. Academic Press, NY.

Yoshikawa Y, *et al.* (2001) Self-assembled pearling structure of long duplex DNA with histone H1. *Europ J Biochem* **268**(9): 2593–2599.

Wulf de P, *et al.* (2003) Hierarchical assembly of the budding yeast kinetochore from multiple subcomplexes. *Genes Dev* **17**: 2902–2921.

Galtsoff PS. (1923) The amoeboid movement of dissociated sponge cells. *Biol Bull* **45**: 153–161.

Müller WEG, *et al.* (1976) Species-specific aggregation factor in sponges IV. *Exp Cell Res* **98**: 31–40.

Gierer A. (1974) Hydra as a model for the development of biological form. *Sci Am* **231**(6): 44–54.

Bonner JT. (1947) Evidence for the formation of cell aggregates by chemotaxis in the development of the slime mold *Dictyostelium discoideum*. *J Exp Zool* **106**: 1–26.

Gerisch G. (1968) Cell aggregation and differentiation in *Dictyostelium*. In: AA Moscona & A Monroy (eds), *Current Topics in Developmental Biology*. Academic Press, NY, USA.

Bonner JT. (1952) *Morphogenesis. An Essay on Development.* Princeton University Press, New Jersey, USA.

Bonner JT. (1959) *The Cellular Slime Molds.* Princeton University Press, Princeton, New Jersey, USA.

Gregg JH, Yu NY. (1975) *Dictyostelium* aggregate-less mutant plasma membranes. *Exp Cell Res* **96**: 283–286.

Kuczmarski ER, Spudich JA. (1980) Phosphorylation of myosin heavy chain *Dictyostelium* regulates self assembly. *20th Ann Meeting Am Soc Cell Biol. J Cell Biol* **87**(2 Part 2): 227A.

de la Roche MA, *et al.* (2002) Signaling pathways regulating *Dictyostelium* myosin II. *J Muscle Res Cell Motility* **23**(7–8): 703–718.

Moscona AA. (1959) Tissues from dissociated cells. *Sci Am* **200**(5): 132–144.

Folkman J, Haudenschild C. (1980) Angiogenesis *in vitro*. *Nature* **288**: 551–556.

Miao M, *et al.* (2005) Structural determinants of cross-linking and hydrophobic domains for self-assembly of elastin-like polypeptides. *Biochemistry* **44**: 14367–14375.

Dubertret L, *et al.* (1987) Les Peaux Artificielles Vivantes. *La Recherche* **18**(185): 149–155.

David I, *et al.* (2002) *The Cambridge Dictionary of Scientists.* Cambridge University Press, Cambridge, UK (Geological time scale).

Dickerson RE. (1980). Cytochrome C and the evolution of energy metabolism. *Sci Am* **242**(3): 136–153.

Vigers GPA, Lohka MJ. (1991). A distinct vesicle population targets membranes and pore complexes to the nuclear envelope in *Xenopus* eggs. *J Cell Biol* **112**: 545–556.

Minshull J, *et al.* (1989) Translation of cyclin mRNA is necessary for extracts of activated *Xenopus* eggs to enter mitosis. *Cell* **56**: 947–956.

Sources of Illustrations

Part VI

6.1 Lima-de-Faria A. (1995) *Biological Periodicity. Its Molecular Mechanism and Evolutionary Implications.* JAI Press, Greenwich, Connecticut, USA (Table 1, page 279).

6.2 Watson JD. (1976) *Molecular Biology of the Gene.* WA Benjamin, NY, USA (Fig. 4.16, page 108).

6.3 (1) Fraenkel-Conrat H. (1962) *Design and Function at the Threshold of Life: The Viruses.* Academic Press, NY, USA (Fig. 20 as a plate).

(2) Becker WM, *et al.* (2003) *The World of the Cell.* Benjamin Cummings, San Francisco, USA (Fig. 2–20, page 35).

6.4 Wood WB, Edgar RS. (1976) Building a bacterial virus. *Sci Am* **217**(1): 60–74 (Fig. Genetic map, page 64).

6.5 (1) Nomura M. (1973) Assembly of bacterial ribosomes. *Science* **179**: 864–873 (Fig. 1, page 869).

(2) Oudet P, *et al.* (1975) Electron microscopic and biochemical evidence that chromatin structure is a repeating unit. *Cell* **4**: 281–300 (Fig. 8B, page 290).

6.6 (1) Pierantoni U. (1944) Tratado de Zoologia. Editorial Labor, Barcelona, Spain (after Kükenthal and Mathes) (Fig. 286, page 395).

(2) Gierer A. (1974) Hydra as a model for the development of biological form. *Sci Am* **231**(6): 44–54 (Fig. Simple organization of hydra, page 46; Fig. Nematocytes, page 53).

6.7   Sussman M. (1978) *Molekularbiologie und Entwicklung*. Verlag P Parey, Berlin (Fig. 64, page 114). From Sussman M. (1964) *Growth and Development*. Prentice-Hall, NY, USA.

6.8   (1)   Vigers GP, Lohka MJ. (1991) A distinct vesicle population targets membranes and pore complexes to the nuclear envelope in *Xenopus* eggs. *J Cell Biol* **112**: 545–556 (Redrawn from Fig. 9, page 555).

       (2)   Moor H. (1967), In: N Higashi (ed), *The World Through the Electron Microscope, Vol III, Biology*. Jeol Co Japan (Fig. Frozen-etched preparation, page 88).

Where Did the Chromosome Come From and Where Is It Going

125

"Where Do We Come From? What Are We? Where Are We Going?" Paintings Which Represent the Origin of Life and of the Chromosome

One of Paul Gauguin's paintings, is the large 1.39 by 3.75 meters oil which he made in 1898. Among the hundreds of art works that the French artist produced, this one has been considered his testament. The canvas is a long panorama depicting a series of pensive naked women resting among symbolic figures. It starts with infancy and finishes with death.

Throughout the ages many artists have painted motives that indicated their anxiety in face of the human condition. One of them is the elusive look and smile of Leonardo da Vinci's Mona Lisa which is a concretization of life's injustice and cruelty (Fig. 7.1).

Gauguin with his brilliant intellect took a step further. He wrote, black on white, that what he intended with this painting was to ask, in concrete terms, what was the origin and the future of life and mankind. In a letter dated February 1898 he described, to Daniel de Monfried, the emotional effort that led him to paint: "Where do we come from? What are we? Where are we going" (Chipp, 1968).

During Gauguin's life time (1848–1903) there was no knowledge of chromosomes, but a few decades later the molecular organization of the chromosome became the center piece of life's origin.

Fig. 7.1   A face expressing life's complexity.

Leonardo da Vinci (1452–1519) was a highly gifted person with the ability to engage in a rainbow of activities extending from engineering to painting. However, he was plagued by personal problems. Life had not been kind to him. The piercing eyes of Lisa Gherardini (wife of Francesco Giocondo) known as Mona Lisa or Gioconda, seem to understand life's trail. They are accompanied by a smile that expresses the cruelty and injustice which are part of the human condition.

Fig. 7.2   RNA synthesized in the test tube for the first time and the origin of the chromosome.

Severo Ochoa (1905–1993), together with Marianne Grunberg-Manago, announced in 1955 the discovery of a new enzyme system capable of synthesizing RNA in the test tube, a crucial step in the understanding of the origin of the genetic code and of life. This finding, together with his numerous contributions to biochemistry, gained for Ochoa the 1959 Nobel Prize. To celebrate his 70th birthday a collection of scientific papers was published. The Spanish artist Salvador Dali (1904–1989), who had several scientists as his friends, contributed with a painting which was used as the book's cover. In it he depicted his version of the assembly of atoms and molecules at the dawn of chromosome formation.

The Spanish artist Salvador Dali, who belonged to a later generation (1904–1989), was strongly influenced by the development of physics and molecular biology during his time. In a book dedicated to the Spanish biochemist and nobel laureate Severo Ochoa (1905–1993), Dali wrote, in 1975, that science had guided his art. In a large painting, dedicated to this molecular biologist, he located nucleic acids, with their atoms, at the center of his canvas symbolizing the chromosome (Kornberg *et al.*, 1976) (Fig. 7.2).

If one leaves the painters and considers the musicians, a score that could have well been written to celebrate the birth of the chromosome is "The Rite of Spring", by the Russian musician Igor Stravinsky (1882–1971). In it the force of life, like the current of a spring river, overflows everywhere and cannot be contained.

126

The Origin of the Cell and of the Chromosome Are Not Known

Living organisms seemed to be so complex that initially many biologists believed that life could not be reduced to chemical analysis. To their astonishment Wöhler (1828) reported the first chemical synthesis of an organic molecule. He had obtained urea starting from inorganic materials. Later other compounds were produced, but it was only in 1953 that Miller synthesized amino acids, which are the constituents of proteins. An electric discharge was applied to a mixture of gases believed to represent the atmospheric composition of the early Earth. To the chemist's surprise the products of the experiment were non-random. He did not obtain a mixture of various types of organic molecules but rather a small number of amino acids and urea (Fig. 7.3). Since then the most significant result has been the synthesis of RNA in the test tube by Grunberg-Manago and Ochoa (1955).

At present one tends to think that RNA, which is such a versatile molecule, has been a main particicpant in the origin of life. DNA is considered to be a late comer in evolution (Fig. 7.4).

Although some of the basic processes of protein and nucleic acid assembly have been elucidated, the exact molecular processes involved in the emergence of the cell and especially of the chromosome remain elusive (Oparin, 1961; Fox, 1973; Schimmel and

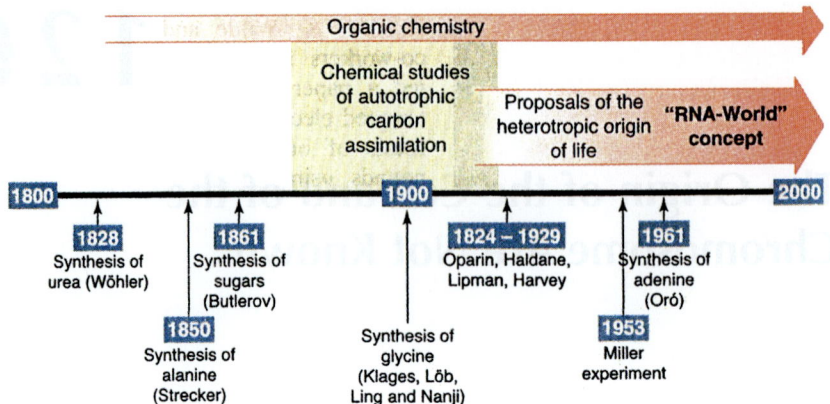

Fig. 7.3 Highlights in organic synthesis in conjunction with proposals on the origin of life.

Following the pioneer work of F. Wöhler (1828) that led to the synthesis of urea, a series of other organic compounds were obtained in the laboratory. But it was the Miller experiment (1953) that opened a new era by obtaining synthetic amino acids. Later work led to the synthesis of RNA (1955) and adenine (1961) in the test tube. At the turn of the 20th century (1900) scientists favoured an explanation of life's origin based on its primordial organisms being endowed with a plant-like (autotrophic) metabolism. This would permit them to use carbon dioxide as their source of cellular carbon. By the 1920s this idea was superseded by the heterotrophic origin of life, proposed mainly by Oparin, according to which life resulted from a series of events involving the synthesis and accumulation of organic compounds into primordial forms. This has led lately to a third view, based on an "RNA World concept", in which RNA would have played a major role as the initial molecule.

Kelley, 2000; Bada and Lazcano, 2003). We simply do not know what atomic interactions led to the self-assembly processes which put together the structural proteins, the histones, the RNAs and the DNAs into a coherently organized organelle which is called the chromosome.

Nucleosomes, the basic constituents of chromosomes, are formed by the self-assembly of DNA with different histones (Oudet *et al.*, 1975) and the interactions between these macromolecules have been established at the atomic level (Luger *et al.*, 1997) but much remains to be learned concerning the atomic processes which shaped the chromosome as a functional unit.

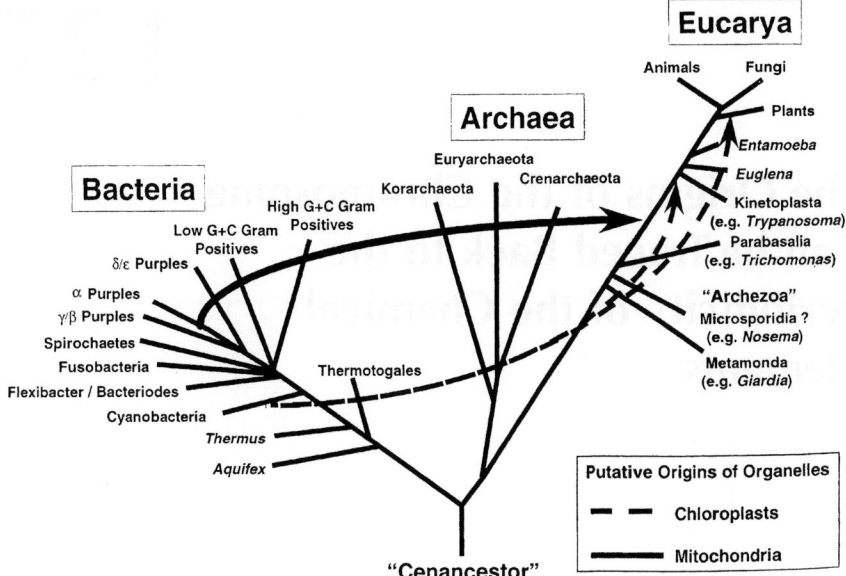

Fig. 7.4   Schematic drawing of a universal tree of life.

This phylogenetic tree is based on the evolution of ribosomal RNA genes. It shows the relative positions of evolutionary groups of the domains Bacteria, Archaea and Eucarya. The putative bacterial origins of the organelles: chloroplasts and mitochondria are indicated by dashed and thick arrows respectively. They start in the Bacteria and finish on the tree branch that leads to plants and animals (Eucarya). As stressed by the authors most of these relationships are open questions.

127

The Origins of the Chromosome Can be Traced Back to the Periodicity of the Chemical Elements

The properties of the chemical elements are by no means arbitrary, since they depend on the structure of the atom, and vary with the atomic number in an ordered way. The significant feature is that there is a periodic recurrence of characteristic properties.

Until 1817 randomness was thought to characterize the chemical elements. When order started to be perceived they were placed in groups of three, and afterwards in groups of seven. Subsequently, the similarities of properties were so evident as to allow predictions to be made concerning missing elements. This was of course the contribution of the Russian chemist Mendeleev who, in 1869, established a relationship between the atomic weights of the elements and their physical and chemical properties. In fact he predicted the existence of six elements that had not yet been discovered. Three (scandium, gallium, and germanium) were soon found while the other three were discovered later. They had the properties predicted by Mendeleev. But it was not until 1922 that Bohr interpreted the Periodic Table in terms of the electronic structure of the atoms (Pauling, 1949). More than a century elapsed before a coherent explanation of elemental order was made available. It became quite apparent that the mechanism beyond

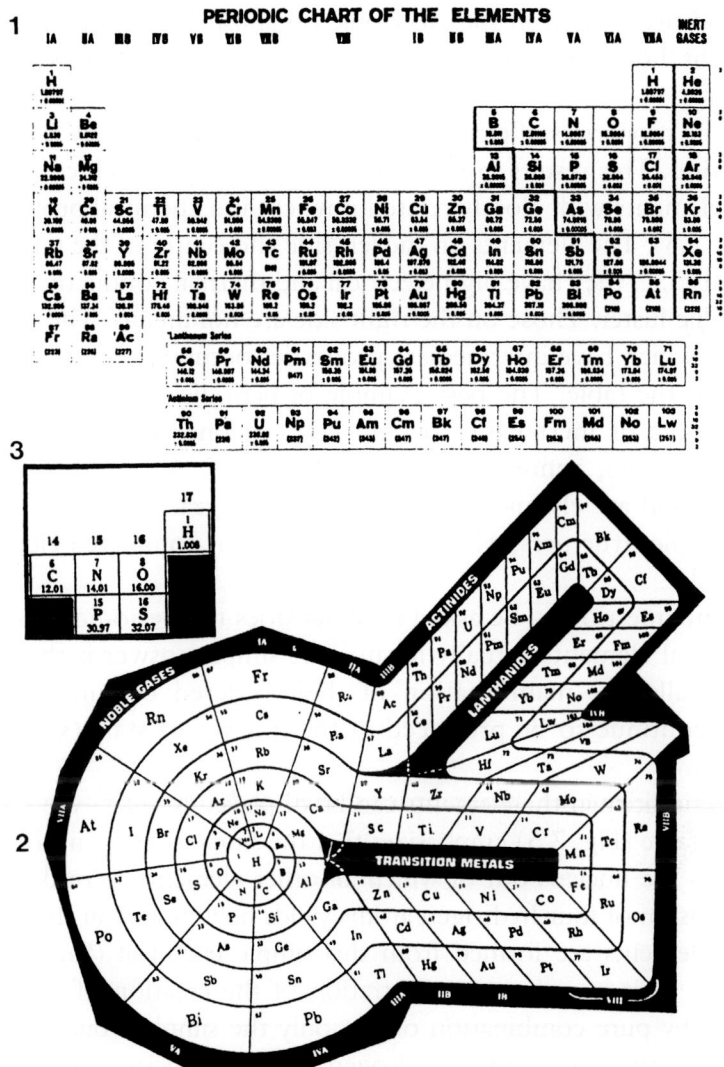

Fig. 7.5   The periodic chart of elements has been drawn in over 700 different forms.

1. In the coventional type of Table which aligns the elements in rows, hydrogen (H) is located both at the extreme upper left and at the extreme upper right, an example of the several exceptional situations that characterize the periodic system. 2. Another form of the chart, based on a spiral, emphasizes the central position of hydrogen which is at the origin of all other elements. 3. The inset between the two forms of the Table, shows the unique position that the atomic constituents of the chromosome (H, C, N, O, P and S) have in the chemical periodicity.

periodicity resided in the electronic organization of the atom. At present, however, much remains to be discovered about this atomic order.

The properties of elements change in a systematic way through a period. This results in the formation of groups which consist of the elements with closely related physical and chemical properties (Fig. 7.5). The elements on the left side and center of the table are metals, i.e., they have high electrical and thermal conductivity and metallic luster. Those on the right side are nonmetals. The metalloids, which have intermediate properties occupy a diagonal position in the table. The best example of periodicity is provided by the noble or inert gases: helium (He), neon (Ne), argon (Ar), krypton (Kr), xenon (Xe), and radon (Rn). All build a single group and are chemically inert due to the great stability of their electronic structures, which results from their completed electron orbitals.

The question to be asked is: What does the periodicity of the chemical elements actually mean? The simple answer is that it is essentially the manifestation of their ordered evolution since they continue to be produced in the interior of stars (Graham, 1996).

Periodicity in the elements can be displayed in many ways. The spiral table (Fig. 7.5) emphasizes that the simplest element, hydrogen, is the nucleus out of which all other elements were derived. The fusion of four hydrogen atoms yields helium and all the other 105 elements are formed from this simple chemical (Sanderson, 1967). This is a clear demonstration of the creation of complex forms by pure combination of not only the simplest but also the same component, namely hydrogen. Such simplicity was possible because the order followed was laid down by discrete energy levels. Thus far, 106 chemical elements have been identified and these give rise to a little over 1,500 variants.

The picture which emerges is that similarity in chemical behavior results from similarity in electronic structure. The elements of each group have usually the same outer electronic configuration, it

is the electrons of the outer shell that determine the basic properties (Asimov, 1992; Masterton and Hurley, 1993).

Not just one, but a whole series of properties show periodicity.
There are no less than 10 periodic trends. This is why the table has
proven to be so important in chemistry.

128

Anomalies Exist at the Level of Chemical Periodicity, but the Alternatives Are Already Limited

No phenomenon reveals total regularity; exceptions and anomalies are known to exist in the most ordered systems. One tends to think that such deviations are characteristic of higher levels of complexity, such as the biological, but they are found already at the atom level. The idea is usually conveyed that the Periodic Table is extremely regular, having no exceptions or deviations. Nothing could be less accurate.

1) To start with, the length of the periods may vary considerably. There are among other sequences, two short periods of eight elements each, two long periods of 18 elements, followed by a very long period of 32 elements (Greenwood and Earnshaw, 1989). This means that already at the chemical level the periods have different lengths.

2) Hydrogen can be located at the beginning of the table over lithium (Li) but it can also be placed, at the other extreme, at the side of helium (He).

3) Helium (He) is an element with an anomalous position in the table. The same is true of europium (Eu) and ytterbium (Yb) that belong to the lanthanide series. Other irregularities include chromium (Cr) and copper (Cu) atoms that carry 1 rather that 2 electrons in their 4s orbitals (Jaffe, 1988).

4)   More than 700 forms of the periodic table have been proposed, each one emphasizing different types of relations between the elements (Mazurs, 1974). Nearly all modern chemistry is based on the periodicity, being a commentary to its implications.

5)   One should realize that every phenomenon is a dynamic process. Although order implies high rigidity, a certain margin of plasticity is present which results in novel and sometimes exceptional solutions, or else no evolution could have taken place. However, it may be noted that already at this level not all kinds of alternatives are allowed.

129

The Unique Position in the Periodic Table of the Atoms Used in the Construction of the Cell and the Chromosome — So Far There is No Evidence that Matter Suddenly Changed Its Laws When the Chromosome Emerged

The question that then arises is what kind of atoms were initially employed in the construction of the chromosome.

Surprisingly there was a preference in atomic combination when the cell and the chromosome were built. As early as 1945, Frey-Wyssling noted that the chemical elements obligatory for cell construction and function were only 25, having a regular distribution in the Periodic Table. All were light elements with an atomic number of less that 35, building mainly a line from carbon to argon (Fig. 7.6). But when it came to the chromosome the restriction became still more evident.

The constituents of chromosomes, i.e. the nucleic acids and proteins, consist of atoms which occupy a particular location in the table. DNA and RNA are built only by: hydrogen, oxygen, carbon, nitrogen and phosphorus (H, O, C, N and P). The polypeptides of proteins consist only of H, O, C, N and S (sulfur). These atoms

Human body		Seawater		Earth's crust	
H	63	H	66	O	47
O	25.5	O	33	Si	28
C	9.5	Cl	0.33	Al	7.9
N	1.4	Na	0.28	Fe	4.5
Ca	0.31	Mg	0.033	Ca	3.5
P	0.22	S	0.017	Na	2.5
Cl	0.03	Ca	0.006	K	2.5
K	0.06	K	0.006	Mg	2.2
S	0.05	C	0.0014	Ti	0.46
Na	0.03	Br	0.0005	H	0.22
Mg	0.01			C	0.19
All others < 0.01		All others < 0.1		All others < 0.1	

REIHEN	0	I	I	II	IV	V	VI	VI	VII	0
1. PERIODE		H								He
2. PERIODE	He	Li	Be	B	C	N	O		F		Ne
3. PERIODE	Ne	Na	Mg	Al	Si	P	S		Cl		(Ar)
4. PERIODE	(Ar)	K / Cu	Ca / Zn	Sc / Ga	Ti / Ge	V / As	Cr / Se	Mn Fe Co Ni / Br			Kr
5. PERIODE	Kr	Rb / Ag	Sr / Cd	Y / In	Zr / Sn	Nb / Sb	Mo / Te	Ma Ru Rh Pd / J			X
6. PERIODE	X	Cs / Au	Ba / Hg	La / Tl	Ce / Pb	Ta / Bi	W / Po	Re Os Ir Pt / -			Em
7. PERIODE	Em	-	Ra	Ac	Th	Pa	Ur				

Fig. 7.6 Non-randomness in the construction of the cell.

1. Comparison of the chemical composition of the human body with that of seawater and the earth's crust. Values are percentages of total numbers of atoms.
2. The chemical elements, essential for cell construction and function, occupy a position in the Periodic Table that is non-random. They tend to build a line that goes from carbon to argon.

have the following characteristics: (1) they are few; (2) four of them are the same in nucleic acids and proteins; (3) the six atoms are not spread over the Periodic Table, but are located only on its right side; (4) they are all non-metals; and (5) they actually occupy a "niche" in the periodic system, all of them having a low number of protons in their atom nucleus. These properties indicate that the periodicity occurring at the atomic level has participated in determining the direction of chromosome organization and function (Lima-de-Faria, 2001).

In its future life, of over 2.5 billion years, the chromosome could not depart from the canalization imposed by these six atoms. During its innumerable replications it could have easily exchanged these atoms for other ones. But such alternatives were blocked by their physico-chemical properties. Wald (1954), a leading specialist of the chemistry of vision, came to the conclusion that the predominance of hydrogen, oxygen, nitrogen and carbon in living organisms "is not at all a matter of chance, but is the inevitable result of certain fundamental properties of these elements" (Eckert and Randall, 1978). Hence, the real canalizers of life properties are not the molecules, nor the atoms, but particles, such as the electrons that inhabit the outer shells of the elements. Trying to bring selection to the electron level is to jump outside the domain of exact science.

Summarizing, to produce the chromosome the cell: (1) Used a non-random process in which it utilized only 6 out of the 106 chemical elements. (2) It also employed only such atoms that had electronic properties that imparted a stability lasting for billions of years. (3) No novel atoms were added to the Periodic Table or new electronic configurations introduced when the chromosome emerged as an independent cell unit.

Thus, so far, there is no evidence that matter suddenly changed its laws, or its evolutionary pattern, when the chromosome appeared on life's horizon.

130

Evolutionary Decisions Which Were Made Before DNA Arrived on the Scene

Evolution did not start with the cell, the chromosome or with DNA. Unless this is realized one becomes trapped in an interpretation of biological evolution that will not allow its full understanding.

Before DNA arrived on the cellular scene several basic processes had been directing evolution for a long time.

1) Symmetries are patent in most living organisms. From where did they come? Not from DNA.

The left-handed and right-handed symmetry that we find in the human body is present already in elementary particles such as the neutrino, which can be either left-handed or right-handed. Subsequently this condition was transmitted, during the evolution of matter, since it is present in the carbon atom and mineral crystals, in which the two opposite forms also occur. It further appears in amino acids and DNA, which is also left-handed (Z form) and right-handed (A and B forms). As living organisms emerged this type of symmetry could not be erased. Left-handed and right-handed forms occur along the whole evolutionary process from invertebrates to humans (Fig. 7.7).

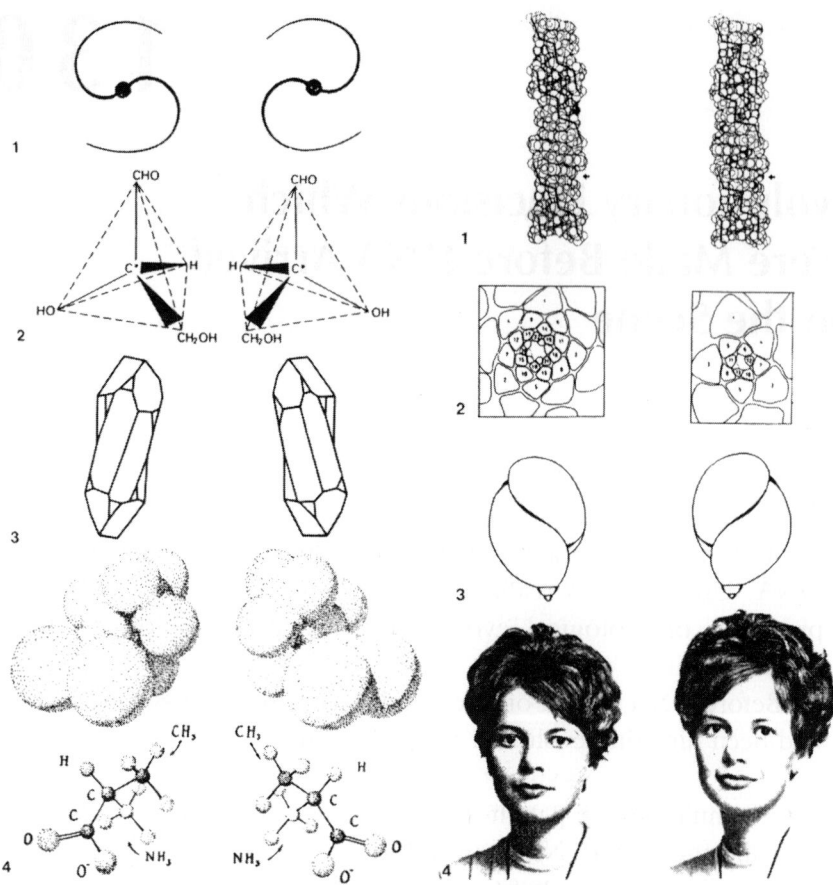

Fig. 7.7   The uninterrupted transfer of symmetries from the dawn of matter to today's organisms. Occurrence of left-handed and right-handed structures from galaxies to humans.

Left column (1) Spiral galaxies of the S-type (left-handed) and Z-type (right-handed). (2) Carbon atom bonds in left-handed and right-handed glyceraldehyde. (3) The mineral quartz in the left and right forms. (4) The amino acid alanine in the left and right form.

Right column (1) DNA in its left (Z configuration) and right forms. (2) Sections through plant shoots (conifer *Araucaria excelsa*) showing spiral divergence left- and right-handed. (3) Shells of the snail *Limnaea* showing left and right orientation. (4) The identical twins Monica and Gerd. Monica is left-handed and has the forelock to the right. Gerd is right-handed and has the forelock to the left. They are mirror images of each other.

Other types, like the twofold, threefold, fourfold, fivefold and sixfold symmetries were present in minerals and quasicrystals (which display fivefold symmetry) before DNA and the chromosome appeared on earth. They were transferred intact from the inorganic world and became later an obligatory component of the plant and animal construction. These symmetries are most evident in flowers where twofold (or bilateral), threefold, fourfold, fivefold and sixfold patterns occur in most plant families. Humans could not depart from the bilateral symmetry. The plants as well as the animals became prisoners of this rigid order (Lima-de-Faria, 1988, 1995, 1997).

But is the chromosome, and its DNA, not involved in the formation of symmetries? Yes, they participate, but only indirectly. They can choose between different mineral alternatives but they have not been able to create new ones. What emerged with the birth of the neutrino is what you find today and what the minerals and quasicrystals were able to organize is also what you find at present in living organisms.

In the plant *Linaria vulgaris* (common toadflax) the bilateral symmetry of its flowers may be transformed into a fivefold symmetry, when one of its genes is modified by a chemical process called methylation. The same phenomenon occurs in the snapdragon (*Antirrhinum majus*) (Cubas *et al.*, 1999) (Fig. 7.8). It may be noted that the DNA bases of the chromosome know nothing about symmetry. This mutation does not even involve a change in the DNA bases, as is usually the case, but is due to a secondary chemical event. The actual transformation of the flower shape is the product of interactions taking place among molecules during the growth of the plant. These events are far removed from the DNA which is locked in the nucleus of the plant's cells.

Hence, the symmetries found in living organisms do not represent any innovation of the gene or the chromosome and are not an immediate product of DNA activity.

2) One usually speaks of "the gene for hemoglobin" or "the gene for chlorophyll" but such expressions are inaccurate

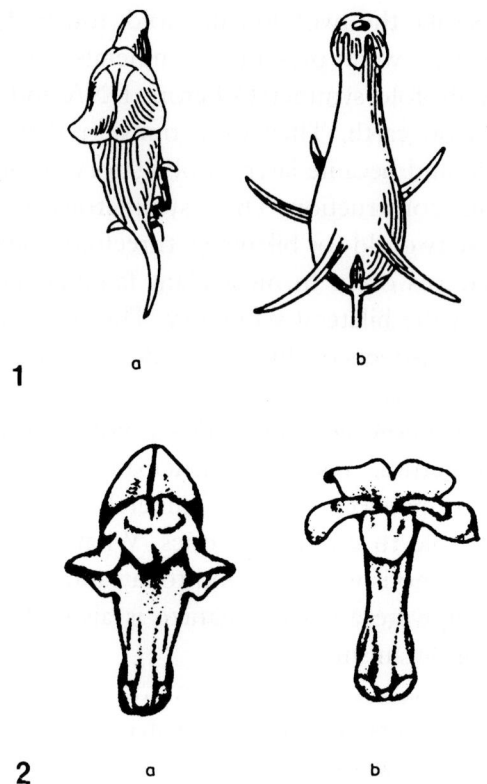

Fig. 7.8   Symmetry changes in flowers.

1. Normal bilateral one-spurred flower of *Linaria vulgaris* (common toadflax) (a) and the five-spurred mutant with fivefold symmetry which is seen also in the upper part of the flower (b).
2. Normal bilateral flower of *Antirrhinum majus* (snapdragon) (a) and the mutant with fivefold symmetry (b).

because they give the false idea that DNA is the main source of the functional ability of these macromolecules.

Hemoglobin contains four iron atoms and chlorophyll one magnesium atom. Along its bases, DNA has no intercalated iron or magnesium atoms. These metals are totally foreign to DNA and this macromolecule lacks the means of creating them anew. The metal ions are collected by the cell from the outer inorganic environment and are later captured by the molecular assemblies that had their remote origin in the DNA of the chromosome.

Two aspects are significant. First, the iron, or the magnesium, as they are integrated into the new molecular edifice do not change their original properties. Second, the information received from DNA in the form of polypeptide chains (in the case of hemoglobin) and specific molecular configurations (as in chlorophyll), is not the carrier of the main function of the macromolecule. The capacity to carry oxygen, which is the primary function of hemoglobin, is located in the four iron atoms that were incorporated (Fig. 3.7). This catalytic activity occurred in ferric ions before iron appeared in hemoglobin (Calvin, 1983). The case is the same with chlorophyll which usually occurs associated with a protein. It is the magnesium porphyrin complex which is responsible for the main function in photosynthesis, one of the crucial events in the history of plant life.

Other macromolecules have incorporated cobalt (vitamin B_{12}) and zinc (zinc finger proteins). Thus, this critical part of evolution is foreign to the base sequences of DNA and the chromosome.

The decisions that shaped the patterns and functions of living organisms such as the symmetries and the capture of metal ions into active molecular edifices, are features that go back to the dawn of matter's evolution.

DNA is not the ruler of these events but a midwife that participates in their perpetuation.

131

The Role of DNA in Heredity is Not as Powerful as We Tend to Believe

There is another aspect of evolution that reveals that DNA is not as important in shaping the hereditary process as we tend to believe.

It is general knowledge that DNA carries the genetic code written on the sequence of its bases which leads to the production of complementary RNAs, well defined proteins and other molecules. But from the DNA molecule to the organism's final shape there are a plethora of intermediate molecular processes that can alter radically the original message written in the code.

Two of these alterations may be mentioned:

1) One of the most significant discoveries of the 1980s was RNA editing. Its significance for the understanding of the evolution of the chromosome and of the cell has not yet been fully realized. RNA editing actually means that the RNA in a single stroke changed the message that it received from the DNA by being spliced in a different way. Hence one does not need a mutation to produce a different protein, and more important, one does not need to change the DNA molecule to create a new function. Moreover, it is not necessary to have a long period of time to create a new molecular edifice. This event modified radically what had before been considered one of the basic processes of evolution. The ability to evolve did not

reside in DNA alone and did not necessarily reside in its capacity to mutate. RNA by itself, when its chemistry found it appropriate, changed the evolutionary scenario and did it instantaneously.

Benne *et al.* discovered the phenomenon of RNA editing in a trypanosome (protozoa) in 1986. The genetic information changed at the level of the RNA. The coding sequence of the RNA differed from the sequence of the DNA from which it was transcribed. The protein coded became different. In the mitochondria of trypanosomes, several bases are systematically added or deleted from the messenger RNA. Four years later it turned out that RNA editing was the result of RNA molecules, which guided the splicing process, functioning as molecular midwives (Blum *et al.*, 1990). Most important was that the RNA splicing occurred *in vitro* in the absence of proteins (Cech and Bass, 1986).

The same phenomenon also occurred in mammals in which different tissues, such as intestine and liver, produced different proteins derived from the same original RNA. The evolutionary modification was caused by a simple change of three bases of the RNA which affected termination and resulted in a much smaller protein being produced (Smith and Sowden, 1996).

One of the most interesting properties of the alternative splicing of RNA is the reversibility of this process (Smith *et al.*, 1989). The order of the RNA segments can be easily reversed, leading to the production of a protein which had not been available for some time.

The decision to make a different protein is inherent in the RNA sequences. As such, the protein emerges without being necessarily correlated with the phylogenetic position of the organism. Two organisms may look very different, and as a consequence may belong to quite different classes or phyla, yet if they happen to have a few genes in common, the old protein may suddenly resurface, creating an unexpected periodical event. The convergence of biological structures or functions, so often observed during evolution, has been usually attributed to the exposure of organisms to similar environments. It may be instead due to the periodic emergence of the same type of proteins (Lima-de-Faria, 1995).

2)   Genome imprinting is another phenomenon that alters gene function. It has been defined as the transcription of a gene, being dependent on its parental origin. This event affects both the expression and the transmission, of particular genes, leading to a violation of Mendel's laws (Villena *et al.*, 2000). In humans this differential expression of paternally and maternally derived genes is effected by long-range mechanisms (Allshire and Bickmore, 2000). An example is Huntington's disease. If the gene is inherited from the father the symptoms show up in adolescence, if inherited from the mother they appear in middle age. This parental origin effect occurs along the biological scale from ferns and flowering plants to invertebrates, birds and mammals.

The role of DNA in heredity is less powerful than one tends to believe, since basic molecular decisions are made independently of its initial message, allowing for evolution to follow unexpected directions and to create subtle variations.

132

The Whole Human Genome May be Packed into a Single Chromosome

It is true that the chromosome follows its undisturbed path dictated by the laws of physics and chemistry. But we may be able to change it appreciably within limits.

As mentioned above, different species of the deer muntjac have 3 and 23 chromosomes in their germ cells. However, the action of their genes results in nearly identical individuals. When we have discovered the rules of chromosome organization that allow such a transformation, we will be in a position to produce a human being with only 3 (instead of 23) chromosomes which is the present number of the human germ cells (46 in somatic tissues). We may even be able to pack all the human genetic information into a single chromosome, as it happens in the ant *Myrmecia*. In these insects a single chromosome produces an animal nearly indistinguishable from another one with 32 chromosomes. At present the molecular mechanisms, involved in the aggregation as well as the dispersal of chromosome units, escape us (Figs. 5.8 and 5.10).

133

Where is the Chromosome Going?

Predictions in science do not represent visions seen through a crystal ball. Rather, they are the outcome of the assembly of data that have been obtained following a certain coherent scheme.

Prediction is a limited process. One simply uses pieces of information that are already available and combines them in a new way. The view, thus put forward, reflects the evidence and technology available at a given time. Obviously a simple answer is not in sight.

Pregnant questions can be asked only when there is solid knowledge available which can be used as a starting point. After 100 years of chromosome research much remains to be elucidated. It is mainly the behavior of the chromosome as a functional unit which escapes us at present (Lima-de-Faria, 2003).

It is only now, following the sequencing of the bases of the DNA of the chromosome complement of several living organisms, including humans, that a series of well-defined questions can be asked. But there are so many black holes in the chromosome's firmament that the results that are pouring in all the time appear as surprises. What seems evident is that the chromosome is a multifaceted structure which so far has withheld from us many of its innermost treasures. Moreover, it has confused us with its antithetical properties.

On one hand, in its relationship to gravity, magnetism, randomness and selection, the chromosome appears mainly as an independent structure. It seems to follow its own trajectory using an internal untamed capacity for innovation. On the other hand, the chromosome appears as a prisoner of the initial atomic construction which canalized its behavior by the process of self-assembly. This has resulted in a rigid pattern that conserved its organization and gene functions for millions of years.

It is this antithetical nature that makes its behavior to look foolish and so difficult to predict.

At present information is accumulating that emphasizes this bewildering situation. Johnson *et al.* (2006) studied the distribution of chromatin sites accessible to transcription, which were localized along the chromosomes of the worm *C. elegans* and concluded that "flexibility and constraint" dominated the structural pattern.

134

Physics is Still an Underdeveloped Science, but It May Hold the Key to the Understanding of Chromosome Behavior

During the past century physics has focused on three main areas of research.

Firstly, the discovery of a large population of elementary and sub-elementary particles which includes leptons and quarks. To this were added particles found in anti-matter such as the positron, the anti-proton and the anti-neutron.

Secondly, four forces have been considered to pervade nature. In order of their increasing strength they are: the gravitational interaction, the weak interaction responsible for radioactivity, the electromagnetic interaction, and the strong quark-binding interaction. The four have been considered the manifestation of but one universal interaction. A great effort has been made to unify the different forces into a coherent synthesis, but the goal remains elusive. This may be due to the fact that gravity pervades all matter and as such cannot be isolated and analyzed *per se*. Moreover, its extreme weakness may add to this difficulty.

Thirdly, astrophysics developed into an independent discipline, due to the great effort employed in understanding the origin and formation of stars, as well as other celestial bodies. In this area new

Fig. 7.9 Physics, like any other science, has a long way to go.

Physics seems to be a highly developed science because it has, in the last years, achieved such startling results, but a close look reveals an intellectual edifice loaded with contradictions, dubious assumptions, uncertain conclusions and above all a vast area of phenomena to be unraveled. As the American physicist R. Feynman put it "I think I can safely say that nobody understands quantum mechanics" (Hey and Walters 2003). The present state of physics is compared with the unfinished tower painted by Bruegel the Elder in 1563 (detail) (Kunsthistorisches Museum, Vienna, Austria).

observations are constantly being added which furnish primary clues of an expanding universe.

This has been a tremendous achievement but physics continues to face an enormous body of data to unravel (Fig. 7.9).

The new century, that has just started, may see physics moving in a reverse direction. After most particles have been sorted out, a new task will emerge. One will search for the laws that direct the combination of the elementary particles as they assemble into more complex units. This will be a period in which one will begin to understand the mechanism responsible for self-assembly. One has to find out whether the rules governing the combination of quarks are the same as those that direct the association of nucleic acids and proteins into viruses, as well as those that obliged the planets to build the solar system.

At the same time physics is probing the imaging of atoms and of elementary particles, an endeavor that will allow to better define some of their still unknown properties. In the 1960s macromolecules, such as DNA, were photographed using the electron microscope. This seemed to be the limit. But in the 1990s single atoms were made visible using more powerful instruments and the sharpness with which they are seen increases all the time (Sugimoto *et al.*, 2007) (Fig. 7.10). But the effort proceeds. The electron is now the

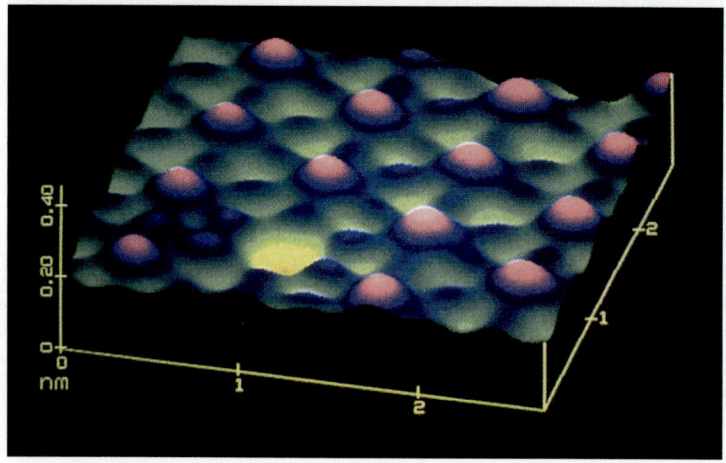

Fig. 7.10   Single atoms seen under the electron microscope.

On one hand, physics has a vast area to discover, but on the other hand it is advancing at great speed. In the 1960s macromolecules, such as DNA, were for the first time photographed with the electron microscope. That seemed to be the limit. But since the 1990s single atoms have been equally photographed using more powerful instruments. The preparation shows iodine atoms which are bonded to each other, but one is missing, leaving a gaping hole in the surface (yellow spot) (nm = nanometer = 10^{-9} meter) (Courtesy of Fran Heyl).

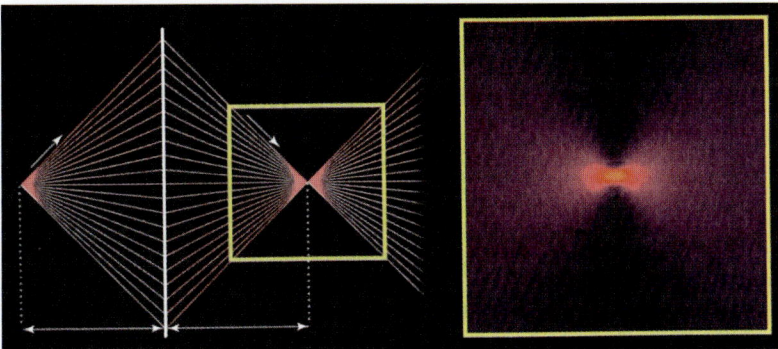

Fig. 7.11   Focusing electrons.

An attempt has been made to focus electrons by using graphene as a lens. This material consists of single sheets of carbon atoms and these have allowed to focus electrons to extremely fine points. (Left) Light formerly diverging from a point source is set in reverse and converges back to a point. (Right) A computer simulation of electron charge density on graphene showing similar focusing.

next target. Graphene, that consists of single sheets of carbon atoms, has been used as a lens to focuse electrons and measure their charge density (Cheianov *et al.*, 2007) (Fig. 7.11).

Another effort is a deeper penetration into the construction of matter. The European Center for Nuclear Research (CERN) located in Geneva, Switzerland, which has been for decades the laboratory of sub-elementary particle research, is now building a machine, the International Linear Collider, that initially will be 31 kilometers long and hurl electrons and positrons with energies of 500 billion electron volts. This collider may be later extended to 50 kilometers having a power of trillion electron volts. The physicists' hope is to discover new laws and forms of matter.

Development is rapid, and it is expected that, soon, self-assembly will be the common ground, on which theoretical physicists and molecular biologists will meet. As a consequence the mechanism of biological evolution will start to unfold, since the data available at present strongly indicate that evolution is a phenomenon inherent to matter (Lima-de-Faria, 1988).

Looking into the future one thing seems most probable. Several generations of biologists and experimental physicists will be kept busy in their efforts to unravel the chromosome's tightly entangled atomic interactions.

In one hundred years from now our picture of the chromosome may have changed so radically that the present view may seem archaic and naive, but that is the expected result of an expanding scientific endeavor. The chromosome has only confessed a fraction of the secrets it holds lifting only a portion of its molecular veil.

References

Part VII

Chipp HB. (1968) *Theories of Modern Art.* University California Press, Berkeley, USA.

Kornberg A, *et al.* (eds) (1976) *Reflections on Biochemistry in Honour of Severo Ochoa*, pp. 445. Pergamon Press, Oxford, UK.

Grunberg-Manago M, Ochoa S. (1955) Enzymatic synthesis and breakdown of polynucleotides — polynucleotide phosphorylase. *J Am Chem Soc* **77**(11): 3165–3166.

Oparin AI. (1961) *Life, its Nature, Origin and Development.* Oliver and Boyd, Edinburgh, UK.

Fox SW. (1973) Origin of the cell: experiments and premises. *Naturwissenschaften* **60**: 359–368.

Schimmel P, Kelley SO. (2000) Exiting an RNA world. *Nat Struct Biol* **7**: 5–7.

Bada JL, Lazcano A. (2003) Prebiotic soup — revisiting the Miller experiment. *Science* **300**: 745–746.

Oudet P, *et al.* (1975) Electron microscopic and biochemical evidence that chromatin structure is a repeating unit. *Cell* **4**: 281–300.

Luger K, *et al.* (1997) Crystal structure of the nucleosome core particle at 2.8Å resolution. *Nature* **389**: 251–260.

Pauling L. (1949) *General Chemistry.* Freeman and Co, San Francisco, USA.

Graham JA. (1996) Time scales and stellar evolution. *Carnegie Institution of Washington. Year Book* **95**: 50–56.

Sanderson RT. (1967) *Inorganic Chemistry.* Reinhold Pub Co, NY, USA.

Asimov I. (1992) *Atom: Journey Across the Subatomic Cosmos*. Truman Talley Books, Plume, NY, USA.

Masterton WL, Hurley CN. (1993) *Chemistry. Principles and Reactions*. Saunders College Publ, Philadelphia, USA.

Greenwood NN, Earnshaw A. (1989) *Chemistry of the Elements*. Pergamon Press, Oxford, UK.

Jaffe HW. (1988) *Introduction to Crystal Chemistry*. Cambridge University Press, Cambridge, UK.

Mazurs EG. (1974) *Graphic Representation of the Periodic System During One Hundred Years*. University Alabama Press, Birmingham, USA.

Frey-Wyssling A. (1945). *Ernährung und Stoffwechsel der Pflanzen*. Zürich, Switzerland.

Lima-de-Faria A. (2001) Genetic mechanisms involved in the periodicity of flight. *Caryologia* **54**: 189–208.

Wald G. (1954) The origin of life. *Sci Am* **191**(2): 44–53.

Eckert R, Randall D. (1978) *Animal Physiology*. Freeman and Co, San Francisco, USA.

Lima-de-Faria A. (1988) *Evolution Without Selection. Form and Function by Autoevolution*. Elsevier, Amsterdam, NY.

Lima-de-Faria A. (1995) *Biological Periodicity. Its Molecular Mechanism and Evolutionary Implications*. JAI Press, Greenwich, Connecticut, USA and London.

Lima-de-Faria A. (1997) The atomic basis of biological symmetry and periodicity. *BioSystems* **43**: 115–135.

Cubas P, *et al*. (1999) An epigenetic mutation responsible for natural variation in floral symmetry. *Nature* **401**: 157–161.

Calvin M. (1983) The path of carbon: from stratosphere to cell. In: K Downey, RW Voellmy, F Ahmad, & J Schultz (eds), *Advances in Gene Technology: Molecular Genetics of Plants and Animals*, pp. 1–35. Miami Winter Symposia, Vol. 20. Academic Press, NY.

Benne R, *et al*. (1986) Major transcript of the frameshifted cox-II gene from trypanosome mitochondria contains four nucleotides that are not encoded in the DNA. *Cell* **46**(6): 819–826.

Blum B, *et al*. (1990) A model for RNA editing in kinetoplastid mitochondria: "guide" RNA molecules transcribed from maxicircle DNA provide the edited information. *Cell* **60**(2): 189–198.

Cech TR, Bass BL. (1986) Biological catalysis by RNA. *Ann Rev Biochem* **55**: 599–630.

Smith HC, Sowden MP. (1996) Base-modification mRNA editing through deamination — the good, the bad and the unregulated. *Trends Genet* **12**: 418–424.

Smith CWJ, *et al.* (1989) Alternative splicing in the control of gene expression. *Ann Rev Genet* **23**: 527–577.

Lima-de-Faria A. (1995) *Biological Periodicity. Its Molecular Mechanism and Evolutionary Implications.* JAI Press, Greenwich, Connecticut, USA and London.

Villena FPM de, *et al.* (2000) Natural selection and the function of genome imprinting: beyond the silenced minority. *TIG* **16**: 573–578.

Allshire R, Bickmore W. (2000) Pausing for thought on the boundaries of imprinting. *Cell* **102**: 705–708.

Lima-de-Faria A. (2003) *One Hundred Years of Chromosome Research and What Remains to be Learned.* Kluwer Academic Publishers, Dordrecht, London; 2004 Springer, Berlin, NY.

Johnson SM, *et al.* (2006) Flexibility and constraint in the nucleosome core landscape of *Caenorhabditis elegans* chromatin. *Genome Res* **16**: 1505–1516.

Sugimoto Y, *et al.* (2007) Chemical identification of individual surface atoms by atomic force microscopy. *Nature* **446**: 64–67.

Cheianov VV, *et al.* (2007) The focusing of electron flow and a Veselago lens in graphene p-n junctions. *Science* **315**: 1252–1255.

Hey T, Walters P. (2003) *The New Quantum Universe.* Cambridge University Press, Cambridge, UK (cited on Fig. 7.10).

Sources of Illustrations

Part VII

7.1   Mettais V. (1997) *Your Visit to the Louvre*. Art Lys, Paris, France (Fig. on back cover).

7.2   Kornberg A, *et al.* (eds) (1976) *Reflections on Biochemistry in Honour of Severo Ochoa*. Pergamon Press, Oxford, UK (Fig. on cover of book).

7.3   Bada JL, Lazcano A. (2003) Prebiotic soup — revisiting the Miller experiment. *Science* **300**: 745–746 (Diagram on page 745).

7.4   Brown JR, *et al.* (2002) Horizontal gene transfer and the universal tree of life. In: M Syvanen & CI Kado (eds) *Horizontal Gene Transfer*. Academic Press, NY, USA (Fig. 26.1, page 306).

7.5   (1)   Sanderson RT. (1967) *Inorganic Chemistry*. Reinhold Pub Co, NY, USA (Table on page 14) (Based on chart published by Fisher Scientific Co).

         (2)   Benfy T. (1964) Spiral periodic chart. In: T Benfy (ed), *Chemistry*. American Chemical Society, Washington, DC **37**(6): 14 (Chart on page 14).

         (3)   Lima-de-Faria A. (2001) Genetic mechanisms involved in the periodicity of flight. *Caryologia* **54**(3): 189–208 (Fig. 15, page 204).

7.6   (1)   Anonymous. (1972) *Biology: An Appreciation of Life*. CRM Books, Del Mar, CA, USA (Table).

         (2)   Frey-Wyssling A. (1945) *Ernährung und Stoffwechsel der Pflanzen*, Zürich, Switzerland (Fig. Periodic Table).

7.7   Lima-de-Faria A. (1995) *Biological Periodicity. Its Molecular Mechanism and Evolutionary Implications.* JAI Press, Greenwich, Connecticut, USA (Fig. 1, page 122; Fig. 2, page 123).

7.8   (1) and (2). Gustafsson Å. (1979) Linnaeus' peloria: the history of a monster. *Theor Appl Genet* **54**: 241–248 (Fig. 2, page 243; Fig. 4, page 245).

7.9   Hagen RM, Hagen R. (1995). *What Great Paintings Say. Old Masters in Detail*, Vol. 1, Taschen (Fig. on page 56).

7.10   von Baeyer HC. (1992) *Taming the Atom. The Emergence of the Visible Microworld.* Random House, NY (Plate following page 102).

7.11   Pendry JB. (2007) Negative refraction for electrons? *Science* **315**: 1226–1227 (Fig. on page 1226, adapted from Fig. 2, page 1253 of Cheianov VV, *et al.* (2007) The focusing of electron flow and a Veselago lens in graphene p-n junctions. *Science* **315**: 1252–1255).

Simplified Glossary

CENTROMERE — A region of the chromosome involved in its active mobility. The microtubules of the spindle fibers attach to its surface during cell division. A replicated chromosome consists of two chromatids joined at a region situated on both sides of the centromere which usually appears as a thin constriction. The chromatids move to opposite poles with the participation of spindle fibers.

CROSSING OVER — The exchange of genetic material between homologous chromosomes, i.e. those received from the mother and father. The exchange occurs during the chromosome pairing which takes place at the early stages of meiosis and results in new combinations of genes.

CYTOPLASM — The substance within the plasma membrane of a cell that does not include the nucleus.

DNA — Deoxyribonucleic acid, one of the two forms of nucleic acid in living cells. It is the genetic material for all cellular lifeforms and many viruses. DNA forms a double helix that is held together by hydrogen bonds between specific pairs of bases. Each strand in the double helix is complementary to its partner strand in terms of its base sequence.

DNA REPAIR — The biochemical processes that correct mutations arising from replication errors and the effects of mutagenic agents.

DNA SEQUENCING — The technique for determining the order of nucleotides in a DNA molecule. A nucleotide consists of a base, a sugar and a phosphoric acid group.

ENHANCERS — Sequences of nucleotides that increase the transcription of other genes. They act by increasing the number of RNA polymerase II molecules transcribing genes located distantly.

EXON — A segment of DNA that is included in the transcript of a gene and survives processing becoming a part of the final RNA message.

GENE — A DNA segment (or RNA in the case of some viruses) containing biological information and hence coding for an RNA and a polypeptide molecule. The DNA may code solely for RNAs which may be of various types and lengths.

GENETIC CODE — The rules that determine which triplet of nucleotides, codes for each amino acid during protein synthesis. The consecutive nucleotide triplets (codons) of DNA and RNA specify the sequence of amino acids in protein synthesis.

GENOME — The total of all genes coding for proteins, as well as for RNAs, together with all other DNA sequences carried by the chromosomes of the complement of an organism which have other functions.

HOMEOBOX — A sequence of about 180 base pairs near the end of certain homeotic genes. Homeobox proteins regulate the transcription and translation of genes during the development of the organism.

INTRON — A segment of DNA that is transcribed into RNA but is subsequently removed from within the transcript and rapidly degraded.

LINE — Abbreviation of "long interspersed nuclear element". A genome-wide repeat often with transposable activity, 6 to 7 thousand base pairs long.

MEIOSIS — The type of division by which the chromosome number of the organism is halved, the result being the formation of sexual cells (or gametes) with half of the chromosomes (haploid value) present in the cells of the body (somatic or diploid number). It comprises two successive nuclear and cellular divisions with only one round of DNA replication. The chromosomes received from the mother and those from the father pair along their whole length (pachytene stage) and contract subsequently becoming tightly coiled. At the first anaphase there is a separation of maternal from paternal chromosome material with the exception of where exchanges have occurred between the two partners (crossing over).

METHYLATION — The methylation of nucleic acids is due to the addition of a methylgroup ($-CH_3$) to DNA.

MITOSIS — The nuclear division which generally consists of four phases: prophase, metaphase, anaphase and telophase. During prophase the chromosomes become visible within the nucleus. Each chromosome is longitudinally double consisting of two chromatids. Subsequently, the nuclear envelope breaks down ejecting the chromosomes into the cytoplasm. At metaphase the chromosomes are located within the spindle fibers arranging themselves in the equatorial region. During anaphase the two chromatids move into opposite poles. As telophase approaches, the spindle disappears and the nuclear envelope is reconstructed encircling the two groups of offspring chromosomes. Next the cytoplasm divides itself into two parts, the result being the formation of two daughter cells.

MUTATION — Structural, or chemical change, of a gene (nucleotide sequence) which is transmitted to the progeny.

NUCLEAR ENVELOPE — Envelope delimiting the nucleus composed of two membranes traversed by ring-shaped pores.

NUCLEAR SAP — The molecular mass which fills the nucleus and in which the chromosomes are immersed.

NUCLEOLUS — An RNA-rich spherical body associated with a specific chromosomal segment, the nucleolus organizer, which contains the ribosomal RNA genes. The nucleolus is composed of the products of these genes, their associated proteins and enzymes.

NUCLEOSOMES — A beadlike structure of eukaryotic chromosomes consisting of a core of eight histone molecules wrapped by a DNA segment of about 150 base pairs and separated from the adjacent nucleosomes by a DNA sequence of about 50 base pairs.

NUCLEOTIDE — One of the monomeric units from which DNA or RNA polymers are constructed. They consist of a purine or pyrimidine base, a sugar, and a phosphoric acid group.

PSEUDOGENE — A gene similar in its DNA sequence to another known gene but which, due to additions or deletions, has become non-functional.

REPLICATION — During DNA replication the two strands of the duplex molecule separate. The enzyme DNA polymerase then adds complementary nucleotides to the two emerging daughter strands.

RNA — Ribonucleic acid, one of the two forms of nucleic acid in living cells. The genetic material of some viruses. RNA molecules tend to be single stranded.

Ribosomal RNA — The RNA molecules that are components of ribosomes.

Messenger RNA — The transcript of a protein-coding gene.

Transfer RNA — A small RNA molecule that acts as an adaptor during translation and is responsible for decoding the genetic code.

Interfering RNAs — Short double-stranded RNAs that participate in silencing complexes.

MicroRNAs — RNAs with a length of 21 to 22 nucleotides. They regulate gene expression by binding to complementary RNAs.

Small RNAs — A class of short RNA molecules involved in the splicing of introns and other molecular events.

RNA EDITING — A mechanism which modifies the nucleotide compositions of previously formed messenger RNAs by adding or deleting uridine molecules at precise sites within the coding regions of messenger RNAs.

RNA SPLICING — A process that results in the removal of introns and joining of exons in RNAs.

RNA SURVEILLANCE — A form of quality control of RNA in which proteins scrutinize the quality of RNAs before they are exported to the cytoplasm.

SELF-ASSEMBLY — The spontaneous aggregation of multimeric biological structures involving formation of weak chemical bonds between surfaces with complementary shapes. Self-assembly is inevitable, automatic and hierarchic.

SINE — Abbreviation of "short interspersed nuclear element" about 300 base pairs long. It is characterized by the *Alu* sequences found in the human genome.

SOS RESPONSE — An error-prone mechanism of repairing damaged DNA in bacteria by the coordinated induction of several enzymes.

SPLIT GENE — Genes containing coding regions (exons) that are interrupted by noncoding regions (introns). This is the organization found in the genes of most higher organisms.

TELOMERE — A specialized DNA sequence found at the ends of linear chromosomes.

TRANSCRIPTION — The synthesis of an RNA copy of a gene. It consists of the formation of an RNA molecule upon a DNA template by complementary base pairing, mediated by RNA polymerase.

TRANSLATION — The synthesis of a polypeptide, the amino acid sequence of which is determined by the nucleotide sequence of a messenger RNA. The translation occurs in a ribosome and leads to protein synthesis.

TRANSPOSON — A relatively long mobile DNA element that moves in the genome by a mechanism involving DNA synthesis and transposition.

Acknowledgements

Thanks are due to my colleague Professor Dr. Bengt Olle Bengtsson, Head of the Genetics Division, for wise leadership and generous support and to Med. Kand. Johan Essen-Möller for excellent computer work. This work was supported by a grant from the Royal Physiographic Society, Sweden.

Name Index

Subject Index

3% of the total DNA, 29

aberrant molecules, 194
absolute zero temperatures, 73
accessory chromosomes, 28,
    206
accidental rearrangements, 157
accurate segregation, 144
actin, 103
action of gravity is invalidated,
    82
active movement of chromo-
    somes, 88
activin, 230, 265
adaptation, 129
adaptor proteins, 193
addition of a poly(A)tail, 195
adenine, 287, 340
adhesion molecules, 219
Agave americana, 282
aggregation, 212
aggregation process, 318
alcohol, 7
alga *Volvox*, 231
algae, 67, 85, 323, 325
alternating current, 260

Alu sequences, 378
amino acid alanine, 352
amino acids, 339
ammonium dihydrogen
    phosphate, 98
Amoeba, 23, 212, 231
Amoeba proteus, 280
amoebae, 227
amoeboid cells, 319
amoeboid slime mold
    Dictyostelium discoideum,
    318, 319
amount of DNA, 29, 283
amphibian tissues, 218
amphibians, 221
amplification, 201
anaphase, 375
ancestral RNA, 200
ancestral RNA-sequence, 199
animal and human populations,
    120
animal pole, 99
ant *Myrmecia*, 359
ant *Myrmecia pilosula*, 277, 278
anterior-posterior axis, 223
antibodies, 214